TEACHING PRIMARY MATHS

Teaching Primary Maths

Building Teacher Confidence and Student Success

Emma Bird

Published in 2025 by Amba Press, Melbourne, Australia
www.ambapress.com.au

© Emma Bird 2025

All rights reserved. No part of this book may be reproduced or transmitted in any form or by any means, electronic or mechanical, including photocopying, recording or by any information storage and retrieval system, without prior permission in writing from the publisher.

Cover design: Tess McCabe
Internal design: Amba Press
Editor: Andrew Campbell

ISBN: 9781923403321 (pbk)
ISBN: 9781923403338 (ebk)

A catalogue record for this book is available from the National Library of Australia.

Contents

About the Author	1
Acknowledgments	3
Introduction	5

Part 1: Who Am I as a Maths Educator? — 7
1. Beliefs Impact Practice — 9

Part 2: Why Does Mathematics Education Matter? — 27
2. The Current Landscape — 29
3. Critical Numeracy — 49

Part 3: What Affects the Learners in My Mathematics Classroom? — 61
4. Essential Engagement — 63
5. Maths Anxiety — 77

Part 4: How Can I Make a Difference in My Own Classroom? — 87
6. Creativity in Mathematics — 89
7. Curiosity in Mathematics — 109
8. Playful Maths — 129
9. Primary Mathematics Essentials — 149

Conclusion	171
References	175

About the Author

Emma Bird is a passionate educator, dedicated school leader, and lifelong learner, who currently serves as Deputy Principal at a primary school in Queensland. With experience teaching across Prep to Year 9 in both state and private schools, she brings a wealth of classroom teaching and leadership experience to her work.

Emma contributed to the revision of the Version 9 Australian Curriculum as a member of the Queensland Curriculum Reference Group for Mathematics P–10. She also acted as a Critical Friend to the Queensland Curriculum and Assessment Authority, advising on how Version 9 should be implemented and supported across all primary subject areas in Queensland.

Emma holds a Graduate Certificate in Primary Mathematics Education and a Master of Education (Mathematics Education) from Western Sydney University, where she was awarded the Dean's Medal for Academic Excellence. Her work has been published in the *Australian Primary Mathematics Classroom (APMC)* journal and the *Queensland Association of Mathematics Teachers* journal.

Emma enjoys sharing her insights into primary mathematics education and building teacher capacity at professional learning events. She is passionate about empowering teachers to feel confident, capable, and inspired in their roles.

Beyond her professional life, Emma is a wife and mother, balancing her love for education with her love for family. Whether being a leader in a school, supporting teachers in the classroom, or engaging in professional research, she continues to model the creativity, curiosity, dedication, and commitment she hopes to inspire in others.

Acknowledgments

Never in my wildest dreams did I think I would write a book about mathematics! Just a year ago, I had completed my master's degree and was looking forward to having my weekends back... but I couldn't shake the feeling that I needed to share what I had learnt with other teachers.

Writing this book was challenging, frustrating, intimidating, rewarding – and a roller coaster of emotions I didn't expect. The hardest part, without doubt, was sacrificing time with my family while I locked in to get the job done. Yet, after years of researching and learning at university, I was surprised by how naturally the words began to flow.

To my husband, Dean, and our two children, Amelia and Elliot – thank you for supporting me and being endlessly patient as I worked nights and weekends. Your unwavering love means the world to me. I hope this book makes you proud and shows that the time we invested in this journey was worth it.

To my mum, dad, brother, and sister – thank you for always being my cheerleaders and believing in me, no matter what. To my Aunty Elizabeth – thank you for challenging me to grow and for your encouragement and support.

To Deb, Pete, Lianie, and Catherine – your willingness to review my work and provide thoughtful, constructive feedback has helped refine my ideas and make the book clearer for teachers. I am deeply grateful for your time and insight.

Finally, to my publisher, Alicia, and my editor, Andrew – thank you for guiding me through the publishing process with professionalism and care. Your expertise has been invaluable in bringing this book to life.

Introduction

Mathematics is more than counting numbers and multiplication – it is a way of thinking, problem-solving, and making sense of the world. In the primary classroom, the foundations we lay in mathematics can shape how students see themselves as learners, thinkers, and capable problem-solvers for years to come. As educators, we play a vital role in nurturing not just mathematical knowledge, but mathematical confidence, curiosity, and resilience.

This book is designed to support you in that journey. Whether you're new to teaching or bringing years of experience, you'll find practical strategies, current research, and real-world examples aimed at deepening your understanding of effective mathematics instruction. Throughout, we'll explore how to create inclusive, engaging, and conceptually rich learning experiences that help all students thrive.

My hope is that this book empowers you to reflect on your practice, try new approaches, and continue growing as a mathematics teacher. Your work matters deeply – and the impact you have on young learners lasts a lifetime. Let's begin the journey together.

PART 1

WHO AM I AS A MATHS EDUCATOR?

1
Beliefs Impact Practice

Your beliefs and experiences impact your teaching practice. Throughout this book, I am going to ask you to reflect deeply on *what* you do as a maths teacher, *how* you do what you do, and most importantly, *why* you do what you do. In order to dig a bit deeper, I am going to need you to lay any well-developed maths armour down so that you can step out of the safe cocoon you may have built around yourself and start your transformation to become a beautiful (maths teaching) butterfly.

You may be raising your eyebrows in alarm. *Did I pick up the wrong book? I thought this book was about* **maths and teaching***, not about* **me and my feelings***!* Take a deep breath, dear reader. You have come to the right place to learn more about maths teaching, but the complicated reality is that when you teach this subject area, you carry more baggage to the lesson than you may realise (and no, I'm not talking about your trolley of manipulatives!). After studying for a Graduate Certificate of Primary Mathematics Education and then completing my Master of Education, majoring in Mathematics Education, I have come to understand the complexities of mathematics teaching.

My passion is to take your hand and guide you on a journey to understand a bit more about yourself as an educator and a bit more about mathematics education in primary school.

My hope for you is that you gain some new insights and reflect on your *how* and *why* by the end of this book. I believe that teachers need to be curious enough and vulnerable enough to continue to be reflective learners throughout their career. The curriculum will change, your students will change, society will change, the learning environment will change, your colleagues will change. Teaching is a dynamic (and wonderful!) profession, but if you are unable to position yourself as a learner throughout your journey, then adding wobble stools to your classroom will be as difficult as teaching a cat to swim. Don't get me started on analogies for how hard the next curriculum re-write will be for you.

Here's the tricky thing. No one can do this for you. You must be willing to do this work of being a learner yourself. When you position yourself as a learner, some parts of this book may feel challenging. Before we can move ahead through this book, though, we will need to take a journey through the past. Don't worry, I'll go first so that you can rest assured that you are not alone in your experiences. I will begin with my story: the story of a girl who felt mediocre in skill and intimidated by class games – a story that has spanned my lifetime and come full circle.

My story

My memories of primary school are mainly quite fond. I generally had great teachers throughout this part of my schooling journey. When I reflect on my maths experiences in particular, there are three main experiences that stand out in my mind.

In Year 3, I had a strict teacher who taught us for a whole year using only lead pencils, erasers, a ruler, and writing books. We silently sat in rows as we independently copied and completed work from the blackboard. My Year 3 maths experiences included copying many sums from the blackboard with perfectly ruled margins, perfectly lined up columns of numbers, with only vertical algorithms being written and solved. There were no discussions, no questions asked, and definitely no drawing diagrams or explaining a concept in another way. The focus was precision of book work, accuracy of calculations, and the speed of getting maths sums completed quickly.

I remember feeling pressure to be fast and perfect. These lessons were the first moments in my schooling journey when I felt stressed and anxious most days. The irony is, I must have been performing to the teacher's standards well enough, because from this age I was selected to be part of the school's 'Maths Extension' program. Thus began the second, most poignant memory of primary school maths… *The Maths Extension Program*.

The Maths Extension Program ran once a week out of the school's learning support room. The room was filled (read: crammed so full that you had to shuffle sideways to get from one seat to another) with old single wooden desks and chairs. Let's just say that the furniture, layout, and dank, dark room would not have been included in the school's beauty spots tour for prospective parents and students. This 'special' space was where my maths anxiety grew stronger with each week. Tasks that we were expected to endure included many logic puzzles. These puzzles infuriated me, as they made no sense and there were no strategies or skills I could apply to them – I had to think laterally to solve them. Every time I left that room of no more than ten students, I felt more and more like a failure. I was convinced that I was the last one accepted into the program and that I was there to keep the seat warm in the tenth spot.

One of the logic puzzles the teacher loved to give us each year (yep, I was 'chosen' to endure this weekly torture for the next four years) is pictured below in Figure 1. Our challenge was to find the single path around the figure without re-tracing any lines or lifting our pencil off of the page. To make an already anxious student even more stressed, there were times when each of us had to stand and demonstrate how we solved the problem on the blackboard, for everyone in the room to see. This problem haunts me to this day. There were times that I could solve it independently, but when I had to get up and draw how I solved it on the blackboard my mind would go so blank that I could barely remember the shape of the figure I was supposed to draw, let alone the starting point.

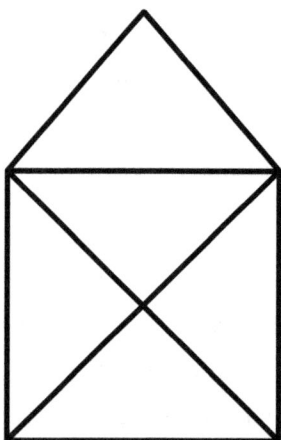

Figure 1. Torturous logic puzzle: recreate the drawing one line at a time, without retracing any line that you have already drawn and without lifting your pencil.

What frustrated me most about this task was that I couldn't understand it. I didn't understand *what* the best path was and *why*. I couldn't comprehend how this related to extending my maths thinking, but I knew that further questions would not be tolerated and that it would be social suicide to unveil myself as the girl who did not understand. It turns out that, 30 years later, I would learn that I was not alone in both my frustration and my desire to want to understand *why*. This is something I will unpack further in Chapter 4.

The third and most embarrassing example of my primary school maths experiences was in Year 7, when it took me almost six months to conquer long division. This time, I had a great teacher who saw that I was motivated to learn, and she gave me tutoring lessons twice a week for over a term, one on one, to help me master this elusive algorithm. I just didn't get it, no matter how hard I tried. Day after day we would practise using long division in our maths lessons, and day after day I was unable to solve this relatively straightforward algorithm.

As an adult now, I realise that the biggest barrier to my mastery of this concept was multiplication. My body would freeze, and my brain would go completely blank (and still sometimes does), when I needed to recall multiplication facts quickly. 'Around the World' (a game that used to be played in classrooms regularly to promote speed of fact recall) did so much damage to my confidence. I would get so nervous playing this game that I would feel nauseous. I would wipe my sweaty palms on my school uniform as my nervous system went into overdrive. As I was being tutored by my teacher, I was often asked to recall the factor of a product to solve the steps of long division. Because of the reinforced idea throughout primary school of the need to regurgitate facts at lightning speed, my brain would shut down, using its survival instincts to protect me.

Fast forward to the beginning of my teaching career almost 20 years ago, and guess what subject I felt most nervous about teaching well? Chances are that you, as a primary school teacher, have also experienced a level of maths anxiety, or know other colleagues who do (more about that in Chapter 5). But here is the decision that I made almost 20 years ago. I chose to position myself as a learner. I chose to get curious. I read books, deep-dived into all of the maths resources I could get my hands on, and asked my colleagues and mentors questions. Before I taught a new concept, I took the time to learn the concept well, understand the developmental sequence (beginning my teaching career using the Year 2 Diagnostic Net was extremely helpful for this), and I thought deeply about the learners in my classroom and their needs. It turns out I was using one of the most effective pedagogical repertoires for teaching maths, called *Pedagogical Content Knowledge* (more on this in Chapter 6).

I used manipulatives, drew diagrams, printed out my own maths story books, designed and cut out my own custom activities, and worked tirelessly to support my students to deeply understand the mathematical concepts I was teaching them. My curiosity about maths grew, and I wanted to develop my understanding further to improve my teaching. I then worked for a consultancy company to develop mathematics resources aligned with each standard of the curriculum, across the elementary years of schooling (the company was based in New York, using the Common Core Standards). This experience challenged me and grew me as the extremely knowledgeable and talented company Director, Dr Liz Irwin, took the time to explain what I did not understand and give me great examples to expand my thinking.

After this time, I began working as a curriculum leader. The Queensland Curriculum and Assessment Authority was calling for expressions of interest from teachers who were willing to contribute their time and knowledge to panels to review Version 8.4 of the Australian Curriculum. I decided to throw my hat into the ring, although I doubted that I would be selected as a panellist. However,

I was selected for the Queensland Advisory Panel for Prep–Year 10 Mathematics, and throughout 2020 and 2021 I read, analysed, and annotated countless documents, and provided as much feedback as I could to help shape the new version of the Australian Curriculum – Version 9.

It was during this time that I realised I was a strong advocate for mathematics education. I decided to go deeper with my curiosity about mathematics teaching by starting my Graduate Certificate in Primary Mathematics. The program was designed by Catherine Attard and was the most fascinating and engaging learning of my career so far. It was run by the most supportive and passionate tutor, Peter MacDonald, at Western Sydney University.

The experience helped me to reconcile my past disillusionment with maths education and to see a bright, positive and promising future for mathematics education in our country. I decided to take a huge personal step by completing my Master of Education, specialising in mathematics education. I chose to dive deeply into the research to learn what I could do to help the students in my school experience maths in engaging and positive ways.

Your story

Which brings me to now. I am passionate about doing maths education well. I want this book to feel like your companion on *your* journey of reflection and learning. I want to help you to grow as a practitioner and to empower you to deliver exceptional mathematics learning that supports students, engages them, and leaves them hungry for more.

If you are willing to come on a journey with me, we will explore the barriers to effective maths education and the steps to take to transform mathematics teaching and learning in our primary schools. Before we begin our journey, it is important that you take the time to reflect on your past. Take time over the coming days or weeks to reflect on your teaching, learning, and past maths experiences. Believe me, this process will help you to uncover so much about why you do what you do and will lay the foundation for you to become a more reflective and effective practitioner.

What were your maths experiences at school?

What positive experiences did you have as a maths learner at school? Take the time to think about this and describe or list these experiences in the space provided. This space is your own journal and has been provided for you to write, sketch, draw diagrams, or use in any other way that suits your needs as a learner.

Did you have negative maths experiences at school?

What were the attitudes of the adults around you?

Describe the attitudes of your teachers, parents, and family towards mathematics.

How did you view yourself as a maths learner?

Describe yourself as a maths learner both as a child and now.

How do you feel about teaching maths?

What are your core values in maths teaching?

What words would you use to describe your maths teaching?

Describe the experiences you facilitate for the learners in your classroom.

What pedagogies do you believe work well in the maths classroom?

PART 2
WHY DOES MATHEMATICS EDUCATION MATTER?

2

The Current Landscape

Before I share some alarming statistics with you, let's talk data. We live in a time when schools and teachers are drowning in data. We collect anecdotal data, observational data, summative reporting data, benchmarking data, wellbeing data, and behavioural data and we participate in large government data-collection assessments like NAPLAN and the Australian Early Development Census (AEDC). Schools now have specific data plans to manage the huge quantities of data, and teachers are overwhelmed by the challenge of interpreting the data. Data is useful in helping us to get a deeper, broader picture of individual students and cohorts. My feeling is that many schools have a strong focus on their internal data and they don't step back to see the bigger, national picture.

In research, data analysis is conducted through triangulation to ensure the accuracy and validity of the data. Triangulation is where three or more data sets are collected independently of each other. When these data sets are compared, contrasted, and analysed, they offer patterns that converge and complement each other, allowing the researcher to triangulate the data. An example of this in your classroom practice is when you observe student responses in an activity and make notes of misconceptions, use PAT maths data, and conduct an assessment aligned with your year level's achievement standard to analyse a student's understanding. Each data set will give you different pieces of the puzzle, and comparing and analysing these data sets side by side can help you to see the bigger picture about your learner.

I have taken the time to explain data triangulation because I want to share some statistics and trends with you. My intent in sharing these trends is to highlight what some of the data tells us about the current trends occurring across our nation. Please be reassured that I know that you, as a teacher, are doing your best in your classroom with your students. I know you have so many conflicting priorities and that it can be hard to hear statistics that report declines in students' proficiency levels because it can feel really personal. I think it is important to value what we do well, but to also step back and see what the big picture is telling us.

AEDC

The 2021 Australian Early Development Census (AEDC) data showed that 17.4% of students are developmentally behind in areas including numeracy (Evidence for Learning, 2023). The AEDC is conducted triennially in the first half of the Foundation Year, across the country. It is the only national data collection of its kind in the world. The data collected measures early childhood development across five domains:

- Physical health and wellbeing
- Social competence
- Emotional maturity

- Language and cognitive skills (school based)
- Communication skills and general knowledge.

The AEDC has been conducted since 2009. In 2021 a total of 1.5 million children were included in the census data (the 2024 data set is yet to be released). The results of this data collection are used to help the government decide where to target funds to support children in vulnerable and disadvantaged communities.

The 2021 AEDC data shows that the percentage of children who were developmentally vulnerable in the domain of language and cognitive skills increased from 6.6% in 2018 to 7.3% in 2021. Further to this, the percentage of vulnerable children in the domain of communication skills and general knowledge increased from 8.2% in 2018 to 8.4% in 2021, with the amount of children on track in this domain decreasing 0.2% to 77.1% in 2021 (Commonwealth of Australia, 2025). I have highlighted these statistics for you to ponder. These are the nationally reported trends reflecting where our youngest primary school students are tracking developmentally, at the start of their schooling journey.

PISA

The Programme for International Student Assessment (PISA), conducted in countries around the world triennially by the Organisation for Economic Co-operation and Development, assesses the skills and knowledge of 15-year-old students in:

- Mathematics
- Reading
- Science.

The tests are designed to 'explore how well students can solve complex problems, think critically and communicate effectively'

(OECD, 2023a). As a nation, Australia first participated in PISA in 2000. The aim of the data is to allow countries to compare their results on an international level and for policy-makers, schools, and teachers to gain insights and learn from the data of other countries.

Although the average results don't show a dramatic decline between each testing period, looking at a graph to see the trends of these results in mathematics over the past 20 years is quite confronting (see Figure 2). There is a long-term decline in our national average mathematics results, with the latest 2022 PISA data showing our 15-year-old students scoring more than 25 points below the 15-year-old students who sat the test in 2002. Most alarmingly, the 2022 PISA data shows that in mathematics our lowest achievers (sitting in the bottom 10% of overall scores) have become weaker.

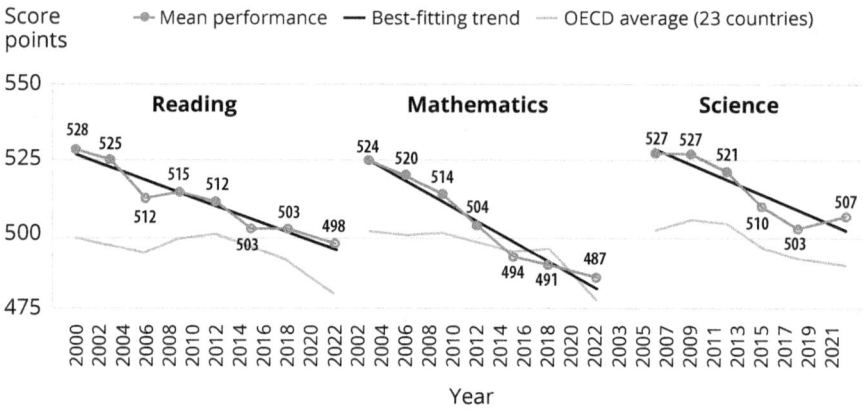

Figure 2. PISA 2022 Results (Volume I and II) – Country Notes: Australia (OECD, 2023b).

However, there is also good news in the latest PISA data: 74% of Australian students met the baseline proficiency score of the mathematics test (which is 6% above the international average). Australia was one of 16 countries (of the 81 countries that participated in

PISA testing in 2022) that saw 12% of students sitting in the top performance levels (the OECD average is 9%).

STEM

The most alarming national trend that has been reported is the decrease in students studying science, technology, engineering, and mathematics at university. Domestic STEM enrolments at Australian universities peaked in 2021 but have since dropped by 3.1% in 2022 and a further 1.4% in 2023. Some areas of domestic university STEM enrolments dropped by an alarming 11% between 2021 and 2023.

Recently, I was having a conversation with a school principal who made the statement 'STEM is dead'. He made this statement because he viewed STEM as an educational trend, not as a skill set that will be valued into the future. STEM careers are essential to our nation. STEM professionals are problem-solvers who work on the world's biggest challenges. STEM skills and professionals put man on the moon. STEM skills and professionals design cars and continue to improve them by making them safer, more economical, and more environmentally friendly. STEM skills and professionals will solve complex global and national problems in the future.

STEM professionals are essential for our nation's ability to innovate and research into the future. Australia's future hinges on those people who are willing to gain university qualifications to learn the complex skills needed for STEM occupations. We need scientists, technology experts, engineers, and mathematicians equipped with higher-level skill sets to solve increasingly complex and novel problems. Equipping our students with a love of STEM subjects – in particular mathematics, which is interwoven across these disciplines – is essential to our nation's future.

Data tells us a story. It is important that we don't only analyse our own school data and that we zoom out to take in the bigger, national picture. Given the statistics shared in just these three data sets, it is clear that there is an overall declining trend in mathematical proficiency from early childhood to high school. The trend is slow at a triennial level, but an increasing gap is being revealed when we compare the longitudinal trends. Along with this trend, we are seeing fewer domestic students enrol in STEM fields at university. Once again, this trend is not significant, but there is a slight decline year on year since 2021 of Australians putting up their hands to be the next scientists, technology experts, engineers, and mathematicians trained at university level, which is a real risk for the future of our nation.

Educational disadvantage in mathematics

As stated on page 32, an alarming trend in PISA data showed that our lowest-achieving students in mathematics have become weaker. Disadvantaged students are less likely to master early maths concepts, which lead to learning more complex mathematics, compounding the cycle of disadvantage over time. This compounding issue occurs because more complex mathematical concepts are difficult to grasp without a firm foundational understanding to build upon.

Low socioeconomic backgrounds

Students from low socioeconomic backgrounds are more likely to be disadvantaged when they start formal schooling. These children are less likely to meet minimum standards in numeracy. As these students reach high school age, they are less likely to participate in STEM subjects, and on average, they are three years behind their peers from high socioeconomic backgrounds. There are more challenges for students from low socioeconomic backgrounds.

Personally, I struggle with these statistics. I attended school in a low socioeconomic area, and these statistics can feel personally confronting. I am passionate about my local area, about its people, and about investing in my community. When I first heard the statistics about the disadvantages faced by students from low socioeconomic areas at university, I was offended by a comment the professor made. He casually remarked, 'For example, in this room, we likely have a majority of students from high socioeconomic areas', with a joking tone and no empathy. People looked around the room at each other, and I felt as though I wanted to shrink into the shadows. I felt exposed and inferior, just because of the suburb I grew up in.

The reality is that students from a low socioeconomic background are more likely to struggle with engagement, have poorer wellbeing, and be less likely to achieve above academic benchmarks. Does this mean all students from low socioeconomic backgrounds are disadvantaged just because of their circumstances or where they live? No, it means that the chances are higher that students from low socioeconomic backgrounds will experience more adversity in schooling than their same-aged peers from high socioeconomic backgrounds.

We need more invested, passionate teachers applying for jobs in these schools to help students from low socioeconomic backgrounds to break out of these statistical cycles. As I will discuss further in Chapter 3, mathematics is the single most powerful predictor of socioeconomic status by middle age, and we, as teachers, have the power to transform the lives of our students both now and into their futures.

Students with disabilities

Students with disabilities and those from disadvantaged backgrounds often face significant challenges in mathematics, which can range from conceptual and emotional difficulties to developmental issues or more complex mathematical conditions like dyscalculia. Addressing these challenges early is crucial, as timely interventions and tailored support lead to better outcomes for students. The sooner these difficulties and misconceptions are identified, the more effectively they can be overcome, helping students build a stronger mathematical foundation.

In Australia, 25.7% of students required educational adjustments due to disability in 2024 (ACARA, 2024), meaning that just over one in four students in Australian classrooms need adjustments made to their learning to allow them to access and participate in education just like their peers. No specific data set for mathematics is collected, but this statistic highlights the need for teachers to be mindful of the learners in their classroom as they have diverse learning needs.

Students with disabilities may encounter challenges in various aspects of mathematics, from basic number sense to more advanced concepts. Teaching mathematics in primary school is akin to building a seven-layer cake – each year, a new layer of understanding is added, with each layer increasing in complexity and expectation. However, the importance of early learning is often overlooked, with some teachers downplaying the foundational skills developed in early years like Prep. This is concerning because for students with disabilities a strong start is essential for future learning to build upon. Each year of primary school mathematics learning sets a critical learning foundation for the next. With each year we build another layer of the learning cake (see Figure 3).

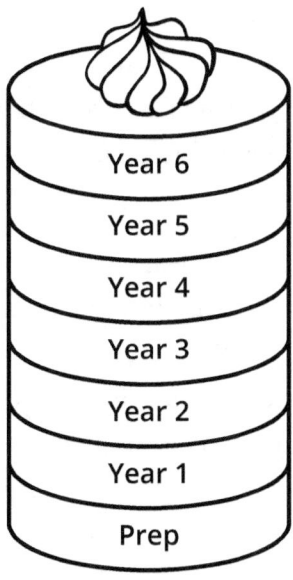

Figure 3. An illustration of mathematics learning in primary school as a seven-layer cake.

For these students, the developmental trajectory can be slower, and they may require additional scaffolding to master key concepts. Without targeted support, they may struggle to connect abstract mathematical ideas to real-world applications. To ensure these students have the best chance of success, teaching strategies must be differentiated to meet their diverse needs. One effective approach is the use of concrete materials to teach mathematical concepts. These hands-on tools reduce cognitive load and provide a solid foundation for deeper conceptual understanding, enabling students to build the necessary skills in a way that aligns with their individual learning needs and supports their ongoing development.

Research shows that students from disadvantaged backgrounds, including those with disabilities, often lack the same educational opportunities as their peers, making it crucial to identify and address their misconceptions early in their mathematical learning.

Teachers must be equipped to observe and assess these students' understanding continually, adjusting their instruction based on formative assessment. This helps ensure that each student is not only supported but challenged in a way that is appropriate for their abilities and learning pace.

For students with disabilities, learning mathematics involves more than simply acquiring mathematical knowledge – it also requires an approach that fosters a positive and inclusive learning environment. This includes addressing the emotional and motivational barriers these students may face, ensuring that they feel confident and capable of succeeding in mathematics, despite the challenges they may encounter.

Ultimately, the goal is to provide equitable access to quality maths instruction for all students, paying close attention to the needs of students with disabilities. By focusing on differentiated teaching methods and fostering a supportive learning environment, we can help these students develop the skills and confidence they need to succeed in mathematics and beyond.

Aboriginal and Torres Strait Islander students

Aboriginal and Torres Strait Islander peoples have long been Australia's first scientists, engineers, and mathematicians. Indigenous peoples developed sophisticated engineering systems, including dams and fish traps, and excelled in water resource management, tool-making, and navigation. These contributions are safeguarded through cultural practices, forming a rich tapestry of knowledge passed down through generations.

In the context of education, the lack of cultural recognition in mainstream teaching models, particularly in mathematics, has led to inequities for Indigenous students. The traditional Eurocentric approach to mathematics education does not account for the cultural capital of students from diverse backgrounds, including Aboriginal and Torres Strait Islander students. By acknowledging how mathematics was historically developed and considering the cultural realities of students, educators can create more inclusive and relevant teaching practices that support all learners.

Aboriginal and Torres Strait Islander students face significant challenges in mathematics education, resulting in lower engagement and poorer outcomes in both further education and employment. These students consistently perform worse in mathematics subjects compared to their non-Indigenous peers.

The work of the Aboriginal and Torres Strait Islander Mathematics Alliance (ATSIMA), led by Dr Chris Matthews, is leading the way in Australian mathematics education, creating new ways of teaching and learning mathematics by connecting mathematics to Aboriginal and Torres Strait Islander histories and cultures. I highly recommend reading the information on ATSIMA's website and attending their professional learning opportunities.

Students with a background of English as an additional or other language

Australian schools are culturally diverse, with students coming from equally diverse linguistic backgrounds. Mathematics education relies heavily on language, as it is crucial for both teaching and expressing mathematical concepts. For English as an additional language/dialect (EAL/D) students, learning maths is particularly challenging because of the complexities and inconsistencies of the English language.

One of the first significant language barriers in mathematics occurs in Prep, when students learn numbers one to ten and then suddenly encounter 'eleven', which represents 1 ten and 1 one. The complexity deepens when we introduce 'twelve', which represents 1 ten and 2 ones. In many other languages, numbers above ten follow a consistent pattern, where the number eleven is directly translated as '1 ten and 1 one', and this pattern continues with numbers like 21 and 31 (see Figure 4 opposite). In English, however, after we introduce 'eleven' and 'twelve', we then use the word 'teen' to describe 1 ten at the front of numbers like thir*teen*, four*teen*, and so on. This language pattern shifts again when we move to numbers containing 2 tens, where we start naming the tens column first, followed by the ones column (e.g., *twenty*-one, *forty*-four), continuing up to ninety-nine. When we consider the complexity of how we use the English language to read numbers, it's remarkable that anyone learns them at all.

To support EAL/D students, particularly students who are beginning to learn English as a second or additional language, teachers should explicitly teach number names using comparative terms from the students' native languages and dialects and establish home–school connections. Additionally, symbolic notation, pictures, and manipulatives can be effective for teaching mathematical concepts.

Mathematical language involves both technical and informal language, and EAL/D students often struggle with worded problems despite having strong conceptual understanding. To address this, teachers must provide clear instructions, technical vocabulary, and visual support, while reducing cognitive load. Teachers should use strategies such as reciprocal teaching, which encourages dialogue and thinking aloud, to foster deeper understanding.

Group work and collaborative learning are effective ways to promote equity and improve communication for EAL/D students, helping them develop both language and cognitive skills. To accommodate diverse student needs, teachers may use Universal Design for

Learning (UDL) principles, ensuring that tasks are accessible to all students. By differentiating instruction and providing various ways for students to demonstrate understanding, teachers can create an inclusive learning environment that supports EAL/D students' success in mathematics.

English Numbers	Chinese Numbers	Translation of Chinese Numbers to English
Nine	九 (jiǔ)	Nine
Ten	十 (shí)	Ten
Eleven	十一 (shí yī)	Ten One
Twelve	十二 (shí èr)	Ten Two
Thirteen	十三 (shí sān)	Ten Three
Fourteen	十四 (shí sì)	Ten Four
Fifteen	十五 (shí wǔ)	Ten Five
Sixteen	十六 (shí liù)	Ten Six
Seventeen	十七 (shí qī)	Ten Seven
Eighteen	十八 (shí bā)	Ten Eight
Nineteen	十九 (shí jiǔ)	Ten Nine
Twenty	二十 (èr shí)	Two Ten
Twenty-One	二十一 (èr shí yī)	Two Ten One
Twenty-Two	二十二 (èr shí èr)	Two Ten Two

Figure 4. English numbers compared to Chinese numbers.

The threat of automation

Automation has benefited our world in many ways. In some hospitals, robots are being used to deliver food and other items to patients, taking pressure off hospital staff who are already very stretched. Automation has made our day-to-day lives easier. Can you imagine handwashing every single item of your clothing, day in and day out? Thanks to the invention of the washing machine, you don't need to! We use dishwashers to wash our dishes, most of us have not had to experience a time when ice in iceboxes needed to be replaced every few days, and newer cars can now park themselves at busy car parks. Yet another example of a more recent innovation that has helped many people in their day-to-day lives is the robotic vacuum cleaner. The most current model empties itself, cleans itself, and does the mopping as well, without you having to lift a finger.

Although automation has helped to improve the quality of our day-to-day lives, it is seen by many as one of the greatest threats to the sustainability of jobs now and into the future. The World Economic Forum's *Future of Jobs Report* (2025) projects that 92 million jobs may be displaced by 2030, mostly due to technological advances and the replacement of human roles by machines. Machines don't need days off, they don't need bathroom and lunch breaks… and they don't need a wage. They are the greatest threat to our students' future job stability and certainty.

Personally, I have seen a decrease in the casual jobs available to high-school-aged students. With the automation of supermarket checkouts, Woolworths, Coles, and Kmart no longer need to employ as many teenagers as checkout operators. The young teens who are willing and capable of working have fewer opportunities to work at these companies because there simply isn't a need for them now. How often have you been to one of those big chain stores in recent times to find that there are few or no checkouts staffed by real people?

Increasingly, jobs are being replaced by machines. This is especially evident where unskilled labour is involved (as at supermarket checkouts). Factories have fewer workers and more machines, some pharmacies are beginning to use robots to fill scripts, and robots are now being used to sort the mail in the Australian postal system. This threat to job security and career longevity also impacts tradespeople, with boilermakers starting to lose their jobs to machines that weld at faster speeds and with greater accuracy.

In a recent discussion with one of my children, we talked about the possibility of graphic design as a future career. I raised my concerns that with the rise of artificial intelligence, the demand for these jobs will likely decrease. All you need to do to generate a bespoke image that reflects exactly the type of picture you have in your mind is describe it to a chatbot, give it feedback to improve the design, and before you know it, you have generated an amazing image (albeit with some warped facial features!).

Future work predictions

The Foundation for Young Australians (2017) predicts that future workers will spend 100% more time on solving problems, over 40% more time on critical thinking, and over 77% more time using maths and science skills each week compared to today's workforce. In addition to this, it is predicted that the children in our primary schools today will have a 'portfolio of careers', holding an average of 17 different jobs over 5 careers throughout their working life. Furthermore, the World Economic Forum's *Future of Jobs Report* (2025) projects that nearly 170 million new jobs will emerge by 2030.

If having 17 different jobs over 5 careers seems far-fetched, let me tell you a bit about my father. After leaving school, my dad learnt his trade as an apprentice shoe repairer. He worked in the inner city and earned a living repairing people's shoes, covering shoes, and making shoes. Shoes were made to last. They were made of quality materials that withstood the test of time, and people bought a pair of shoes with the intention of wearing them, along with only a few other pairs, over the next few years. Because of the quality of the shoe materials and the extent to which people wore them, soles and heels could wear down or become uneven over time. It was my father's job to repair these parts, allowing people to get even more use out of their shoes. Shoes were made so well that when a lady bought a good pair of heels and wanted to change the colour, or the fabric outer layer, she would visit a shoe repairer to re-dye or re-cover them. My father ran his own business as a sole employee. He worked very hard and had an excellent reputation within the community, with many satisfied customers who frequented his business.

In the 1990s, the Australian Federal Government used trade liberalisation, primarily through tariff reductions, to enhance the growth of the Australian economy through integration with the global economy. Cheap shoe stores began to open, shoe manufacturing quality declined, and there was a significant decrease in the cost of shoes due to manufacturing being outsourced to other countries. Seemingly overnight, my family was facing potential bankruptcy. What was once a modestly sustainable business now faced a steep decrease in customers and the need for shoe repairing. Why repair shoes that cost you less than $20?

While still running his shoe repair business, my father diversified and learnt how to cut keys. From this sideways step in his shop, he became interested in locks, selling his business and going back to trade college to become a locksmith. My father is a dedicated worker who rarely has days off. He devoted his time to a locksmithing business for many years. After mastering another trade and discovering a love for learning, my father developed a strong desire to continue learning more. He decided that it was time to pursue the next steps of learning, and he applied to study at university. Another career change was on the horizon.

One Christmas, my father announced that he had applied to get into university to become a teacher and that he had been accepted. We were so excited for him! He worked extremely hard to become a teacher specialising in manual arts and adult education. He is a high school teacher to this day.

There are threads of commonality that came together in my father's career story – trades and teaching students trade skills being the obvious ones. But when my father started his career in 1978, he started with the intention of being a shoe repairer for his working life. He would never have guessed that circumstances beyond his control would take him in the direction of three completely different careers – from shoes, to locks, to teaching.

There are common threads in work life that are transferable across careers. These common threads are called soft skills. Soft skills are a set of traits that are used in the workplace and can include:

- Creativity
- Teamwork
- Self-regulation
- Emotional intelligence
- Problem-solving
- Critical thinking
- Time management
- Communication.

We have already seen an increase in the understanding and explicit integration of these types of skills across schools since the introduction of the Australian Curriculum in 2010, when core elements, named General Capabilities, were introduced as one of the dimensions of the curriculum. General Capabilities embedded within the Australian Curriculum include:

- Critical and creative thinking
- Digital literacy
- Ethical understanding
- Intercultural understanding
- Literacy
- Numeracy
- Personal and social capability.

These General Capabilities encompass what the Australian Curriculum and Assessment Authority calls the knowledge, skills, behaviours, and dispositions for learners of the 21st century. Soft skills can be strengthened through teaching these skills and integrating them into your everyday teaching practice. This is not some educational trend; these are real-life, real-world skills that the students in our classroom today will need to rely upon in their future careers.

CHAPTER SUMMARY

- Data in schools is useful in helping us to understand more about individual students and cohorts, but it is important for us to step back and see the bigger, national picture.

- There is an overall declining trend in mathematical proficiency from early childhood to high school. The decline is slow at a triennial level, but an increasing gap is being revealed when we compare the longitudinal trends.

- Disadvantaged students are less likely to master early maths concepts, creating a compounding cycle of disadvantage over time.

- Although automation has brought many advantages to our society and households, it is a very real threat both now and into the future to job security as more machines are used in place of humans.

- Primary school students of today are predicted to hold an average of 17 different jobs over 5 careers.

- It is important that students develop their soft skills, as these skills are transferrable across jobs and are predicted to be increasingly needed in the future workplace.

3
Critical Numeracy

The words *numeracy* and *mathematics* are often used interchangeably, but it is important that we take a moment to define the two terms. Although they possess commonalities, they also have distinct differences, and we need to ensure that we are using these terms correctly in professional conversation.

The Victorian Government defines numeracy as 'the knowledge, skills, behaviours and dispositions that students need in order to use mathematics in a wide range of situations' (Victorian State Government Department of Education, 2024). The *Cambridge Dictionary* (Cambridge University Press, 2025) describes numeracy as the 'ability to do basic mathematics'. The word *do* is an important one in distinguishing between the terms numeracy and mathematics. In short, numeracy is the *application* of mathematical knowledge and skills. In schools, we aim to develop literate students who will grow to be able to fill in forms, read important information, and communicate orally and through the written word. We equally aim to develop numerate citizens who can use and apply the knowledge and skills that they gain from mathematics learning to accomplish simple daily tasks like telling the time, cooking, shopping, and managing money, up to more complex tasks like planning trips, interpreting timetables, reading maps, and comparing mobile phone plans.

Mathematics is defined as the 'study of numbers, shapes, and space using reason and usually a special system of symbols and rules for organising them' (Cambridge University Press, 2025). Mathematics is a standalone subject taught across all year levels in the Australian Curriculum, where students develop the 'knowledge, skills, procedures and processes in number, algebra, measurement, space, statistics and probability' (Australian Curriculum, 2025). We use mathematics in our everyday lives and apply our mathematical knowledge and skills purposefully using numeracy.

In summary, mathematics is what students learn, and numeracy is the application of this knowledge. Figure 5 opposite is my best attempt to explain this further using a Venn Diagram. Essentially, mathematics and numeracy are interlinked, interdependent, and crucial for everyday life.

Critical social numeracy

Numeracy is deeply connected to equity and social justice. Longitudinal research by Ritchie and Bates (2013) reveals a striking correlation: a child's maths skills at age seven can predict their socioeconomic status at age 42. This finding underlines the far-reaching implications of numeracy development in early education. Numeracy proficiency can open pathways to academic success, career advancement, and economic stability, while its absence can lead to disadvantage.

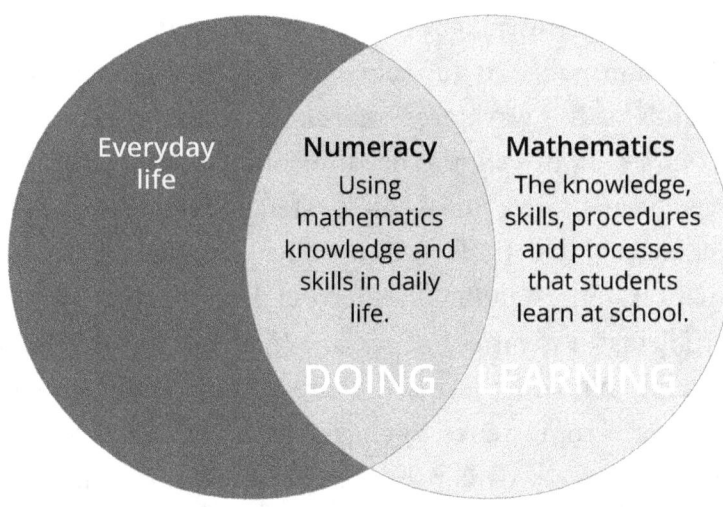

Numeracy	Mathematics
The use of mathematics in everyday life, for example: • telling the time • cooking • shopping • managing money • interpreting timetables • reading maps • comparing mobile phone plans • interpreting graphs in the media • understanding health information.	The learning of mathematical knowledge, skills, and processes, for example: • number • algebra • measurement • space • statistics and probability.
DOING	LEARNING

Figure 5. Numeracy versus mathematics.

In contemporary Australian society, the capacity to engage with numerical information is fundamental to effective citizenship. As the demands of the workforce and daily life become increasingly complex and data-driven, the need for individuals to be numerate has never been more critical. This extends beyond the ability to perform basic calculations – it includes the capability to apply mathematical understanding to interpret data, make decisions, and solve problems in a variety of ways. When we, as teachers, build our students' numeracy skills, we are equipping them with the confidence and competence to engage with mathematical ideas in real-life situations – whether they're managing money, interpreting graphs in the media, or thinking critically about data in society. It is the responsibility of schools and teachers to foster numerate citizens who can participate fully and thoughtfully in democratic and economic life.

Throughout formalised schooling, numeracy is a core component of the Australian Curriculum and encapsulates a student's ability to know and use mathematics purposefully in various contexts (Australian Curriculum, 2024). The fact that numeracy is an embedded element of the Australian Curriculum, as a General Capability, means that every teacher, regardless of the subject that they teach, has a responsibility for developing students' numeracy capabilities. Numeracy is about using mathematics purposefully, whether that's in science experiments, reading timetables, or designing digital games.

The integration of numeracy throughout the curriculum reflects a broader societal vision. Informed, numerate individuals are better equipped to interpret statistics presented to them in advertisements, manage personal finances, understand health information, and participate meaningfully in society. As such, developing critical social numeracy – numeracy that enables thoughtful and responsible citizenship – is not only a goal within the education system but an essential skill for our nation and indeed our world. Students must be prepared to navigate a world that increasingly relies on data, algorithms, and mathematical models. They need to be able to interpret, understand, and analyse information critically to make informed decisions.

Despite this emphasis in the Australian Curriculum, the reality is that trends within our schools eventually impact society. Currently, there are trends of declining mathematical proficiency among our students in measures of standardised performance as assessed by both national (NAPLAN) and international (PISA) data (Cohrssen et al., 2013). Despite these results, the regular feedback I hear from teachers, particularly in primary schools, is that there is a lack of opportunity for professional development in mathematics teaching, with a greater emphasis on improving student literacy.

I am not suggesting that one subject has more worth than the other, but rather that both literacy *and* numeracy are critical areas for students to develop skills in. As a primary school teacher, I consider myself to be a generalist. I advocate for all subject areas that I teach, value each subject in the curriculum, and strive to deliver rich learning experiences for my students that meet their learning needs and help them to grow holistically as individuals. Over the (almost) two decades of my career, I have witnessed an inequitable focus on literacy, which I feel has been to the detriment of developing numeracy skills. For instance, the Queensland Government recently committed $35 million to reading programs (Grace, 2023), while no equivalent investment has been directed toward mathematics-specific initiatives in nearly a decade. The last major funding of this nature was a $183 million investment in 2015 targeting general improvements in literacy and numeracy for NAPLAN performance (Newman & Langbroek, 2014).

This discrepancy reflects an enduring imbalance in educational priorities. While literacy is undeniably crucial, the undervaluing of numeracy contributes to widening gaps in students' numeracy capabilities and may limit their future opportunities. We need to step back and reflect on what both the research and the data trends are telling us; it is time to invest in the development of our students' numeracy capabilities. This responsibility lies with individual teachers, who need to attend to their own professional learning to develop their mathematics education teaching skills; with schools, which need to strategically invest in mathematics education through professional learning and support for teachers; and with governments, which need to invest financially to support schools in building teacher knowledge and capacity in primary mathematics education.

It is essential to elevate the value of numeracy as a critical life skill. Developing numeracy means equipping students with the ability to confidently and accurately apply mathematical thinking in everyday situations. This includes understanding interest rates when applying for a loan, comparing mobile phone plans, interpreting health statistics, managing a household budget, or making sense of election data. When we frame numeracy in this way, it becomes clear that it plays a crucial role in informed decision-making and active participation in society. We must treat numeracy as a critical, foundational competency that empowers students throughout their lives.

Early maths skills

Mathematical understanding develops long before formal schooling and is crucially shaped by early experiences in the years before school and throughout the early years of primary education. In fact, research at the University of Pennsylvania found that babies as young as 16 weeks old detected changes in dot patterns when images of two and three dots were shown to them (Sousa, 2015) – demonstrating that mathematical skills like counting are innate human capabilities that are evident from early infancy. Early maths skills (EMS) encompass a broad set of mathematical competencies, including counting, pattern recognition, spatial awareness, measurement, and data handling, which children learn to apply purposefully across various contexts. These competencies form the foundation for later academic success and are, according to a growing body of evidence, the single most powerful predictor of long-term achievement in school (Anders & Rossbach, 2015; Clements et al., 2016; Parlakian, 2022).

The importance of EMS cannot be overstated. Research has repeatedly shown that early mathematical ability surpasses early reading skills as a predictor of later academic achievement. In their comprehensive literature review involving over 285 scholarly sources, Clements et al. (2023) emphasised that 'maths is a core component of cognition' (p. 2), reinforcing its interconnectedness with the development of language, executive function, and social skills. Early exposure to mathematical concepts not only supports learning in mathematics itself but contributes to overall cognitive growth.

EMS typically develop from birth through age seven, a period marked by rapid cognitive development. The Early Years Learning Framework states that EMS developed in pre-school settings are 'understandings about number, patterns, measurement, time, spatial awareness and chance, and data, as well as mathematical thinking, reasoning and counting' (Australian Government, Department of Education, 2022, p. 67). Cohrssen et al. (2013) suggest that EMS are as broad as number, measurement, data, and space; with one-to-one correspondence, cardinality, and stable order considered foundational EMS. Sousa (2015) takes the view that pre-school students should be exposed to concepts in number, geometry and spatial relations, measurement, patterns, and data analysis. EMS are varied and multi-dimensional, with similar threads of conceptual themes agreed upon throughout the literature. For clarity of meaning in this book, early maths skills (EMS) are defined as: *mathematical competencies in counting, patterns, spatial awareness, measurement, and data that students purposefully use in a range of contexts.* This definition encompasses the core elements of mathematics in the Early Years Learning Framework, which informs learning in Australian early learning centres and kindergartens before formalised schooling begins. It also encapsulates the essence of the numeracy General Capability embedded in the Australian Curriculum and is rounded out with key elements outlined by Clements et al. (2023), Cohrssen et al. (2013), and Sousa (2015).

Anders and Rossbach (2015) found that preschool teachers' past experiences with mathematics, their attitudes toward the subject, and their underlying pedagogical beliefs significantly influence their teaching practices. In their study of 221 teachers, those with more positive attitudes and greater enjoyment of maths were more successful in identifying and integrating mathematical concepts. In contrast, teachers who experienced anxiety or held negative perceptions of mathematics were less likely to notice or capitalise on mathematical learning opportunities in everyday activities.

Similarly, Clements and Sarama (2018) argue that persistent myths about young children's mathematical abilities continue to undermine effective teaching. These myths – such as the belief that young children are not ready for abstract thinking or that maths should be delayed until formal schooling – can limit the use of evidence-based teaching strategies. When educators internalise these misconceptions, children are at risk of experiencing maths as a rigid, narrow subject, rather than as a dynamic, meaningful part of their world.

Our beliefs and attitudes towards mathematics education shape the decisions that we make as teachers and impact the learning opportunities that we provide for our students. A great resource that summarises both national and international research and offers practical recommendations for teaching students is the *Evidence for Learning* guidance report (2023). This report reveals that 17.4% of Australian children start school developmentally behind in numeracy, a disadvantage that can persist and widen over time. I recommend that you take the time to read this report, as the information in it is informed, interesting, and practical. Reflect on the information contained in the report and consider what small tweaks you could make to your teaching practice to improve your pedagogy.

Reflect on the information contained in the report and consider what small tweaks you could make to your teaching practice to improve your pedagogy.

The development of EMS is crucial for our students. The work done by teachers in our early primary classrooms forms the critical foundation for further mathematics learning. When I learnt that EMS are *critical* for a student's future success, I had a visceral reaction. I delved deeper into the research and found that a longitudinal study of a cohort of over 18,000 people aged 7–50 conducted between 1965 and 2009 in England, Scotland, and Wales found that mathematical knowledge at seven years old predicts the socioeconomic status of a 42-year-old adult (Ritchie & Bates, 2013). Even though this study was not conducted in Australia, the sheer number of people surveyed over an extensive period in various socioeconomic contexts holds significance for Australians. The study demonstrates, longitudinally, that socioeconomic status is positively impacted by developing early maths skills.

After learning more about the longitudinal research conducted by Ritchie and Bates, and also delving deeper into what mathematics education research says, I decided to spend the better part of two years investigating this theme further. As a result, I completed my Master of Education (Mathematics Education) with a laser focus on the development of early maths skills, which involved undertaking a case study research project with a small group of Prep teachers. I was internally driven by questions that motivated me to dig deeper and learn more, for the good of our students: How is this not common knowledge in our schools? Why do we continue to undervalue the importance of mathematics education in primary school settings? Why are our primary school teachers still feeling so overwhelmed by the demands of teaching mathematics?

Reflect on the longitudinal research findings from Ritchie and Bates. How do these findings make you feel? Do they surprise you?

I am interested to know your thoughts and feelings about this information. When I first learnt how critical it was to develop numeracy skills in early primary and indeed throughout primary schooling as a firm foundation for learning, I felt quite overwhelmed. I felt personally responsible, and it was confronting to learn what the research has found about the outcomes for students later in life in relation to their numeracy skills at age seven. My biggest motivator in writing this book was to communicate this information to you, as a primary school teacher, because I think it is important for you to know the long-term impacts of prioritising the development of early mathematics skills so that you can refine your mathematics teaching to improve outcomes for your students. Throughout this book, we will explore some practical suggestions that you may choose to implement in your classroom that aim to empower your students and improve their engagement and learning in mathematics.

CHAPTER SUMMARY

- *Mathematics* is what students learn (learning), and *numeracy* is the application of this knowledge (doing).

- Mathematics and numeracy are interlinked, interdependent, and crucial for everyday life.

- Numeracy is deeply connected to equity and social justice.

- A child's maths skills at age seven can predict their socioeconomic status at age 42.

- It is the responsibility of schools and teachers to foster numerate citizens who can participate fully and thoughtfully in democratic and economic life.

- Numeracy is a core component of the Australian Curriculum and encapsulates a student's ability to know and use mathematics purposefully in various contexts.

- Early maths skills (EMS) encompass a broad set of mathematical competencies, including counting, pattern recognition, spatial awareness, measurement, and data handling, which children learn to apply purposefully across various contexts.

- Research has repeatedly shown that early mathematical ability surpasses early reading skills as a predictor of later academic achievement (Clements et al., 2023).

PART 3

WHAT AFFECTS THE LEARNERS IN MY MATHEMATICS CLASSROOM?

4

Essential Engagement

Witnessing our students participate in lessons demonstrating a self-driven motivation to learn is one of the ultimate goals for us as teachers. We like to see our students entering our classrooms wanting to learn, but as they saying goes, 'You can lead a horse to water, but you can't make it drink'. We can create what we feel are the most exciting and fun lessons, but students do not always want to engage with what we have planned for them. Engagement is more complex than we realise and involves a few elements that we will explore in more depth throughout this chapter.

Attard (2022) states that engagement is a multi-dimensional construct, where each dimension works with the others, and the combination results in deep engagement with mathematics that can be sustained over time. She defines the dimensions of engagement as cognitive, operative, and affective. Ribeiro et al. (2023) agree that engagement is multi-dimensional, involving cognitive, behavioural, and emotional elements. Hidayatullah et al. (2023) contend that student engagement involves both 'behavioural engagement' (p. 3), through attention, effort, and participation, and 'emotional engagement' (p. 3), involving energy and enthusiasm. For clarity, I define engagement as three dimensions of *cognitive*, *behavioural* and *emotional* engagement encompassing the work of Attard (2022), Hidayatullah et al. (2023), and Ribeiro et al. (2023).

Hidayatullah et al. (2023) found that when students were engaged, enthusiastic, and interested in mathematics, they had a strong, positive belief in their ability to do mathematics, which meant that they were more likely to do well in the subject. Additionally, Connell and Wellborn (1991, as cited in Ribeiro et al., 2023) assert that engagement is 'the expression of motivation via the child's investment in learning and school activities' (p. 2). Conversely, Ribeiro et al. (2023) assert that when students are not engaged, they may exhibit anti-social behaviours where they find participation and following instruction difficult, resulting in negative learning experiences. Engagement in mathematics lessons is therefore essential, as it not only promotes a positive atmosphere for learning but also supports students in viewing themselves as capable learners, encouraging participation and pro-social behaviours.

Engagement in mathematics lessons is critical to fostering motivated and capable learners. While teachers strive to create dynamic and exciting learning experiences, student engagement remains a complex, multi-dimensional construct influenced by cognitive, behavioural, and emotional factors. When students are engaged, they are more likely to feel confident in their mathematical abilities and exhibit enthusiasm, effort, and positive behaviours in the classroom. Conversely, a lack of engagement can lead to disinterest, behavioural issues, and negative learning outcomes. Therefore, understanding and promoting engagement is essential for cultivating a supportive and effective mathematics learning environment.

Elements of engagement

Cognitive engagement   Behavioural engagement   Emotional engagement

Figure 6. An illustration of the three elements of student engagement (Attard, 2022; Hidayatullah et al., 2023; Ribeiro et al., 2023).

Engagement in practice

Throughout 2024, I conducted a research project to explore the experiences of Prep teachers implementing play-based mathematics lessons to enhance early mathematics skills. I worked with three teachers to hear their experiences throughout their careers, I delivered professional learning to them, and they delivered lessons that I had constructed. I had the pleasure of being able to interview these three teachers to hear about both their experiences before and insights after the project. I deeply appreciate the time that these three teachers gave me to do this project. I learnt so much from them, and one of the richest themes that emerged from the project was engagement.

Engagement was a sub-question of my research, but I never set out to investigate engagement as a construct and main theme. However, engagement was one of the strongest themes that emerged during the project. I had avoided using the word engagement in my key research question because how do you measure engagement? Often, we measure engagement in our classrooms through our own perception of the lesson. We measure how engaged students were by their willingness to participate in the lesson, and by how enthusiastic they were about the lesson. Our own feelings on that day can also impact how well we feel students engaged in a lesson. Were we 'feeling it' during the lesson? Were we disrupted by a phone call from the office, by a parent needing to pick up their child early for an appointment, or by a student coming back from their music lesson and having to have what they missed explained to them? All of these things impact the flow of the lesson, and in turn how we feel the lesson went and how well our students were able to maintain focus and enthusiasm with the inevitable interruptions that happen within our day.

I was interested to see clear themes of engagement emerge throughout the research project. As a result of the professional learning I delivered to the teachers and their willingness to be reflective about their teaching and responsive to their learners, I saw marked growth in how the teachers described and identified engagement in their lessons. Figure 7 summarises some of the themes of engagement that the teachers shared with me in one-to-one interviews. In the first interview, I asked the teachers a range of questions that prompted them to describe their learners and how they knew they were engaged in a lesson. In the second interview I asked the same questions, and the teachers drew on what they observed during the lessons I had designed for them.

In both the first and final interviews, the three teachers described engagement in mathematics. In Interview 1, teachers provided examples of engagement across one to two elements (see Figure 7). After delivering the lessons and participating in the professional learning session, they articulated engagement across all three elements with rich detail, illustrating what engagement looked, sounded, and felt like in their mathematics lessons (see Figure 7).

	Cognitive engagement	Behavioural engagement	Emotional engagement
Interview 1		• Participating • Sharing • Following instructions • Being 'on task' • Active participants • Discussions	• Enthusiasm • Keen • Fun • Smiling • Laughing
Interview 2	• Captured • Challenging • Thinking • Writing and recording ideas • Taking ownership • Discussion • Making links • Associating with prior knowledge • Creative • Discovery • Making mistakes • Modifying • Asking questions • Making connections • Peer feedback • Natural extension • Seeking information • Pretending • Experimenting	• Interacting • Sharing • Initiating • Taking the lead • Busy • Quietly listening • Discussion • Talking • Peer collaboration • Active movement • Chatter • Independent	• Interested • Engrossed • Enthusiasm • Persistence • Resilience • Excited • Enjoyment • 'Eyes twinkling' • Eager • Fun

Figure 7. Teacher descriptions of engagement in pre-lesson and post-lesson interviews, categorised against the three elements of student engagement (Attard, 2022; Hidayatullah et al., 2023; Ribeiro et al., 2023).

One teacher defined engagement as enthusiasm, peer interaction, and participation. After the play-based lessons, she noted multi-dimensional engagement, citing examples of students showing confidence, taking turns, and supporting each other. In a challenging lesson, students displayed excitement and frustration. During this lesson, students encouraged each other to persist, and the teacher reassured students that it was 'ok to make mistakes'. This supportive environment aligns with research on mindset that suggests students are likely to both persevere when they are challenged and achieve higher grades when they hold a positive perception of their ability (Yeager & Dweck, 2020).

Initially, another teacher described engagement as active participation and body language. She enhanced lessons with an engaging voice and an imaginary character, which excited students. This teacher described engagement in the second interview as students being immersed, sitting on the edge of their seats, engaging in peer collaboration and feedback, building on each other's ideas, sharing their thinking with the teacher, asking questions, and displaying persistence and resilience, even when they were challenged by one of the lesson tasks. This teacher reported that she did not need to employ behaviour management strategies due to student engagement in the lessons. This aligns with the findings of Ribeiro et al. (2023) who found that if students had a positive perception of mathematics, they were likely to be engaged both behaviourally and emotionally in their learning.

The third teacher showed a significant shift from describing engagement as purely behavioural to incorporating all three elements of engagement in the second interview. She highlighted active participation, where students focused, asked questions, and involved her in their work.

Engagement impacts learning

Engagement has a clear impact on learning. When students are not engaged, we see detrimental effects on their focus, behaviour, learning, and retention of information. One of the most powerful things to experience is a lesson that hits in just the right spot. You see students who are willing to participate (behaviourally engaged), are enthusiastic (emotionally engaged), and ask questions to clarify their understanding (cognitively engaged). It is also a really great learning experience for teachers to experience what it looks, sounds, and feels like when students are not engaged, because you are likely to reflect on the lesson and try to problem-solve to see what went wrong and what you can adjust for next time to ensure more engagement and a more successful lesson.

Teachers can clearly see and feel when students are engaged in a lesson. Although it is not realistic to think that you can make every lesson deeply engaging, successfully accomplishing all three elements of engagement in a lesson, it is great practice to reflect on lessons to evaluate what worked well for the learners in your classroom, in your context, in your current cohort. What works one year with one group of learners may not work well the next year with the next group of learners. As teachers, we learn how to read our students both as individuals and as cohorts. Each student is unique, and each cohort is unique. That's one of the most beautiful things about teaching: the uniqueness of our students makes teaching a dynamic and exciting profession. We need to adapt lessons to our learners, to their interests, strengths, and needs, and of course deliver this learning within the framework of the curriculum.

I feel that learning gets a supercharged feeling when students are engaged on all three levels. There is a vibe in the classroom that has such a special energy. This type of engagement is why most of us do what we do, and it really brings learning to life – there is a 'visceral feeling' in a classroom when students are deeply engaged. Not only is there a positive feeling for teachers, and in the learning that is occurring in the classroom, but engagement also impacts how students view themselves as learners.

When students are able to experience all three levels of cognitive, behavioural, and emotional engagement, they are able to think deeply. They find it easier to understand the content of the lesson, to retain information, and to generalise this learning beyond the specific lesson. I often describe the challenge of teaching students to my staff like there are two islands in a child's brain. One island is the information they learn and the other island is the contexts in which they may use this learning. Our challenge as teachers is not only to help our students to build solid islands of information and learning, but also to support students in building a tightrope that connects the learning to make it more generalisable in another context, then strengthening this to a rope bridge, and eventually helping students to create a more stable and strong bridge that they can easily cross to draw on both what they know and how to apply this knowledge.

Girls and maths

There is evidence to suggest that gender has a role to play in how students experience and feel about mathematics (Boaler, 2022; Sousa, 2015), with one study conducted by Murtagh et al. (2022) finding that play-based mathematics learning has a positive and significant impact on girls. Dr Jo Boaler (2022) stumbled across an interesting juxtaposition between girls and boys and their attitudes towards mathematics. She found that boys felt more connected and happier when there was competition, speed, and they were getting answers correct. Girls desired to know *why* a particular solution worked, *where* it came from, and how it *connected* with other mathematical methods (Boaler, 2022). When gender disparities are removed and equal contributions are expected from both females and males, female students are more likely to engage and perform better (Sousa, 2015). In fact, when students are encouraged to discuss their thinking and think more flexibly about problems, and when teachers set up an environment where there are open-ended possibilities for solutions, males and females perform not just equally but at higher levels than in classrooms that do not adopt this pedagogical approach (Boaler, 2022).

Reflecting on my own experiences of primary school mathematics, which I shared in Chapter 1, I absolutely agree with what the research says. I felt immense anxiety and pressure with games like 'Around the World', and these anxiety-provoking experiences made me feel inferior and incapable. Fast forward over 30 years and I have now completed a Master of Education majoring in mathematics. I enjoy finding out how a mathematical concept is developed, to help students to understand, and why a mathematical concept may be trickier or easier to understand. I read books about maths, listen to podcasts, discuss it with other teachers, try things out with students, reflect on what I observe, and find ways to refine my practice to improve student outcomes. I can honestly say that I enjoy doing this. It brings me great joy to be able to know and understand how mathematics is learnt and to push myself to improve my understanding of the teaching of mathematics.

Considering that girls make up approximately 50% of our students (in a co-educational school), it is important to design lessons that cater to both genders. Using a Universal Design for Learning (UDL) approach provides both boys and girls with equitable opportunities to have positive learning experiences. My favourite analogy for what it means to design learning experiences using a UDL lens is designing buildings with stairs. When designing a building, say a public library, we may choose to design an entrance with stairs. Many people would be able to access the library using the stairs, but people in wheelchairs would not be able to enter this public building. However, we could add a ramp to the building. By adding the ramp, the person in a wheelchair can now easily access the building, as can the parent with a pram, the elderly person using a walker, and the person who has broken their leg and is using crutches. UDL is when we not only have the stairs as the entry, but we make a considered decision to add a ramp during the initial design/planning of the units, lessons, and assessments, helping a targeted group of students and then opening up possibilities for even more students to access the learning.

Using a UDL lens to view mathematics lessons, we need to think about what learning we deliver for our students and how we deliver it. If we deliver mathematics lessons with a focus on speed and competition, we are likely to make most of our girls (approximately 50% of our students) feel more anxious and disconnected from the learning. To build the ramp and allow equitable access for students of all genders, we could encourage learning through fun games that still have an element of competition (with a de-emphasis on speed), but we could also incorporate time for students to reflect on the strategies they employed in the game, giving time for them to ask their peers about their strategies so that all students can reflect on their thinking and refine their strategies for playing the game. Essentially, by allowing students to ask questions, and de-emphasising speed and competition, we open up the learning to be more equitable for all students, regardless of their gender.

Case study

Born in the USA in 1918, Katherine Johnson was a girl who loved to learn and use maths in her everyday life. Katherine loved to count, and she counted everything and anything she came across (steps to cross the road, pews in her church, cutlery on the family table). By age 10, she was accelerated to high school, and by 15, she was accepted into college.

In 1953, Katherine began working for NASA (which at the time was known as the National Advisory Committee for Aeronautics [NACA]) in the role of a 'computer'. Before the invention of the device that we now know as a computer, the role of a 'computer' as a job was to manually perform complex mathematical calculations for scientific or business purposes. Women frequently held the computer roles and completed calculations for male engineers.

By the time Katherine started working at NASA, she had already completed extensive higher degree study, including work to become a research mathematician. Katherine worked on her assigned mathematical problems in her role as computer, but she would ask lots of questions. In an article on the NASA website, summarising the details of Katherine's career and life, Heather Deiss (2020) states that Katherine had a desire to go beyond simply performing calculations and that she asked many questions. Katherine wanted to know 'Why?', 'How?' and 'Why not?' She was not satisfied with simply performing calculations but had a burning desire to know more.

In the movie *Hidden Figures* (Melfi, 2016), the plight of Katherine Johnson is explored in a dramatic retelling of her story. The movie depicts not only Katherine's gift in mathematics but the severe discrimination that she experienced as both a woman and a person of African-American descent. Katherine showed true courage and determination as she used her intellect and creativity and ultimately leveraged her ability to ask 'How?', 'Why?' and 'Why not?' to solve one of the most complex mathematical problems of the time. Katherine's ability to ask questions and see problems from different perspectives was one of the key factors that enabled NASA to win the 'space race' and send man to the moon in 1969.

We need to empower our girls to feel like they can be successful in mathematics – just as successful as the boys. We need to encourage them to ask questions, to see things from a range of perspectives, and to try a range of ways to solve problems. We never know when this mathematical knowledge will be needed to solve a novel problem in an innovative way. Katherine was up against significant oppression, and ultimately her gift as a mathematician, coupled with her ability to ask great questions, allowed her to solve a problem that was baffling some of the brightest mathematical and scientific minds of the time, ensuring that her name went down in history.

CHAPTER SUMMARY

- Engagement is multi-dimensional and includes the dimensions of *cognitive*, *behavioural* and *emotional* engagement.

- Engagement in mathematics lessons is critical to fostering motivated and capable learners.

- When students are engaged, they are more likely to feel confident in their mathematical abilities and exhibit enthusiasm, effort, and positive behaviours in the classroom.

- When students are not engaged, we see detrimental effects on their focus, behaviour, learning, and retention of information.

- While boys enjoy competition and speed in mathematics lessons, girls desire to know *why* a particular solution worked, *where* it came from, and how it is *connected* with other mathematical methods (Boaler, 2022).

- When students are encouraged to discuss their thinking and think more flexibly about problems, and when teachers set up an environment where there are open-ended possibilities for solutions, males and females perform not just equally but at higher levels than in classrooms that do not adopt this pedagogical approach (Boaler, 2022).

5

Maths Anxiety

Before we explore maths anxiety – its history, prevalence, and impact on learning – we must first define the term. 'Mathematics anxiety describes feelings of apprehension, tension or discomfort experienced by many individuals when performing mathematics or in a mathematical context' (Richardson & Suinn, 1972, as cited in Carey et al., 2019).

Maths anxiety is not a new phenomenon; in fact, a study that uncovered signs of stress while completing mathematics questions took place in the 1950s, with Richardson and Suinn being the first researchers to develop a scale to measure maths anxiety in 1972. Zaslavsky goes as far as saying that for the level of logical knowledge that mathematics as a subject requires, it 'inspires more emotion than any other school subject' (Zaslavsky, 1994, p. 5, as cited in Department of Education and Training Victoria, 2020).

Maths anxiety is one of the things that I think makes mathematics as a subject so unique. When I was studying mathematics at university level, almost every person I came across would make comments like:

'I hate maths!', 'Better you than me', 'I'm no good at maths', 'I never liked maths', or 'You must be really smart.' I'm saddened that people have such strong (and mainly negative) reactions to mathematics as adults, especially in general conversation. There are no follow-up questions, no further discussion, just very strong and emotive views of themselves and of me. I am sad for them that they have had such negative experiences when learning mathematics, and that these feelings continue to feel so big for them. We all have different interests; just because one person pursues their interests through further study in one particular subject area does not make them 'smarter' than someone else. I think we can all learn a lot from each other and that we need all sorts of professions, jobs, hobbies, and interests.

Although anxiety can be experienced in other subject areas around tests, performing, or speaking, maths anxiety has been widely researched and has been found to have its own measure of anxiety that is more severe than one may experience in relation to tests or performances for learning and assessment in other subject areas (Dowker et al., 2016). Maths anxiety has been found to include symptoms of increased heart rate and breathing (physiological symptoms), and negative thoughts, worries, or rumination (cognitive symptoms) (Department of Education and Training Victoria, 2020). Maths anxiety is very real.

Maths anxiety in the classroom

Maths anxiety is a common and significant obstacle to learning and can be seen in children as young as five years old (Dowker et al., 2016). According to Boaler (2022), approximately one-third of school students report that they experience aversion and anxiety in mathematics, with up to 50% of adults experiencing maths anxiety. The Department of Education and Training Victoria (2020) states that 25% of 15-year-old students had feelings of helplessness when engaging in a mathematics problem but estimate that up to 17% of people experience maths anxiety.

Maths anxiety impacts teachers as well as students. In a study of 221 early years teachers, Anders and Rossbach (2015) found that when a teacher experienced maths anxiety and negative attitudes towards the subject, the quality of their teaching practice was negatively affected, along with children's learning. Compounding issues of maths anxiety in students and teachers are the persistent myths that surround mathematics in the early years.

Clements and Sarama (2018) explored eight common myths in early maths education. Although conducting a literature review on mathematical myths may not seem rigorous at first glance, on deeper inspection, these common myths have the potential to have a negative effect on students, as evidence-based teaching strategies and pedagogies may be ignored because of educators' falsely held beliefs. Clements and Sarama (2018) contend that widely accepted myths in early maths education have the potential to 'mis-educate' (p. 4) students and potentially limit the application of best practice. It is essential to examine evidence-based mathematics pedagogies to ensure deep, connected and positive learning experiences.

Maths anxiety is evident across pedagogical contexts. DePascale et al. (2023) conducted a study of 106 four-to-five-year-olds in Atlanta, USA, as they had identified a research gap in the examination of maths avoidance and anxiety in play-based contexts. The researchers used the *Child Maths Anxiety Measure* to get baseline data on maths anxiety before students completed a task. Children counted the dots on a card and then pressed the toy button to correspond with this number. The researchers observed the behaviours of the children, particularly when they counted the correct number of dots and the toy did not make a noise to confirm that they were right. DePascale et al. found that children with lower maths anxiety were more likely to explain why the toy did not work with maths-related statements and reasoning, remaining calm and engaged in the task. Although this study was conducted in another country and a specific socio-economic area of high-income family backgrounds, the behaviours of the children in the study highlight the positive impact on children's engagement and resilience in a task when children have low maths anxiety and are presented with a mathematical challenge. Conversely, the study demonstrates the potentially negative consequences for learning when young children are experiencing maths anxiety. Positive and engaging experiences in mathematics are essential for learning.

The impact of maths anxiety on learning

Like any form of anxiety, when a person experiences maths anxiety, their body reacts. It is well documented that physical symptoms such as increased heart rate occur, along with cognitive symptoms like worrying. As discussed in Chapter 1, I myself experience maths anxiety, and when I do so, my brain goes completely blank as I try to apply logical thought to the mathematics problem at hand.

Maths anxiety can affect any learner in the classroom. Dowker et al. (2016) analysed 60 years of research to understand more about what we have learnt about maths anxiety over that period. They summarised various research findings that concluded that maths anxiety may present in the following ways in the classroom:

- Students may experience working memory deficits.
- Students may be slower to complete tasks.
- Students may make more errors when they are experiencing maths anxiety.
- Students may avoid situations and activities that involve maths.

Boaler (2022) argues that students need to view maths as a subject they can think about and make sense of. Furthermore, Clements and Sarama (2018) propose that instead of children seeing mathematics as narrow and characterised by rote learning, teachers need to employ pedagogies that support students to see maths creatively and connectedly. Teachers need to use evidence-based pedagogies in mathematics to unlock the potential for creativity, deep thinking, and joy. Rather than reinforcing negative attitudes and anxiety, positive mathematics experiences that expose mathematics as a 'broad landscape of unexplored puzzles' (Boaler, 2022, p. 34) have the potential to build resilience and confidence in young learners.

Supporting primary school teachers

Throughout the past few years, I have been working to support my school community through the transition to the Australian Version 9 Mathematics Curriculum. As a member of my state curriculum review team during 2020 and 2021, I was privileged to be able to see this curriculum grow, change, and develop over time. Curriculum writing is no easy task – it is complex and nuanced – yet it feels exciting to be part of the process of trying to make meaningful change for students and teachers. There were many changes between Version 8.4 and Version 9 of the Mathematics Curriculum. This sparked a lot of media attention, causing teachers to feel negative about the curriculum before they even had a chance to familiarise themselves with it and use it in practice.

Mindful of the changes to the curriculum and of the evidence around maths anxiety, I worked slowly over a few years to roll out a range of initiatives that were designed to incrementally build teacher confidence and to support them in their mathematical content knowledge. Primary school teachers are generalists; they teach all subject areas and need to have a level of expertise in all subject areas. As a curriculum leader, I need to be mindful of the many expectations teachers have across each subject area and I need to strategically plan so that teachers feel supported, informed, and empowered. First, we embarked on a journey of completing the Mathematical Mindsets professional learning webinars available through YouCubed online, delivered by Jo Boaler. Each teacher received their own dot journal to use as their personal maths journal, to support them in recording and reflecting on their learning in ways that worked best for them.

In the year before our school's familiarisation with the new curriculum, I called for the formation of a Maths Action Group. The Maths Action Group was a team of teachers who worked with me to research, survey staff, students and parents, and form actionable, evidence-based steps to support teachers to begin familiarisation the following year. We developed a bespoke, whole-school framework and an action plan for professional learning and support based on our survey data. As part of professional learning, teachers engaged in explicit learning about the development of particular concepts (e.g., additive thinking). They explored the resources we had, and the purposes of those resources, and we worked collaboratively to build a bank of resource summaries for teachers to use when planning their lessons. They engaged in maths investigations from start to finish and were given time to reflect through discussion and in their maths journals.

The framework was 'soft launched' for six months before our official implementation year, to give teachers the time to use the framework and seek support or clarification. Teachers were armed with practical tools like lesson examples, resources, and mathematics coaching to refine their practice.

Knowing what the research says about maths anxiety, I knew that I had to journey with teachers to support them in their own mindset and feelings about mathematics, both as a learner and as a teacher. We strove to support our teachers to develop their confidence and capacity, and I am so proud of the work they have done. But we are not done yet. This is an area that I, as a leader, need to continue to support, and I am always mindful that 20–30% of my teachers may experience maths anxiety. As a result of this process, here is what I have learnt is best to support teachers:

Slow and steady is best

It is not a sprint, but a marathon. Slow, sustained effort with supports in place for teachers to develop their own knowledge and understanding is essential. We need to allow time for people to adjust to change, respond to their needs, and adjust timelines where necessary. We need to aim to develop quality teaching and learning, by building teacher confidence and professional capacity.

Give plenty of examples

I know that I gain so much from professional learning sessions and always learn something new, but I particularly love to have something tangible to engage with or take with me to help transfer what I have learnt to my context. Be careful of 'death by PowerPoint' and ensure that you give teachers plenty of good-quality examples to use. Examples are where the rubber hits the road and could include books, maths games, infographics, concrete materials, spreadsheets or Padlets (https://padlet.com), with links to further self-directed learning through videos, articles, and quality online resources.

Provide plenty of opportunity for professional learning in a variety of contexts

We need to be really mindful of supporting teachers in both their professional learning and their wellbeing when planning and delivering professional learning in mathematics. Use a wide variety of professional learning to give teachers opportunities to have positive engagement. This may include:

- Professional learning sessions like workshops or information sessions
- Collaborative planning opportunities
- Time for peer feedback to refine assessment tasks
- Coaching with targeted feedback and support
- Revisiting key concepts and ideas in a range of formats
- Giving teachers time to actually play with maths games
- Giving teachers time to explore and analyse the great resources in your school
- Developing collaborative resource repositories which teachers can help to build with the key information needed (at our school this includes the name of the resource, page numbers/weblinks, Content Descriptions, year levels the resource may be most suitable for, and a space for notes where teachers can add further information about the resource that they feel may be useful to their colleagues).

Give teachers their own maths journal

In order to implement whole-school practices like maths journalling, you need to allow teachers time to practise this themselves, to see how other people journal, and to reflect on their own learning. We use an A5 dot journal, and each teacher is given one when they start at our school. They bring this journal to maths professional learning sessions.

CHAPTER SUMMARY

- 'Mathematics anxiety describes feelings of apprehension, tension or discomfort experienced by many individuals when performing mathematics or in a mathematical context' (Richardson & Suinn, 1972, as cited in Carey et al., 2019).

- Maths anxiety has been found to include symptoms of increased heart rate and breathing (physiological symptoms), and negative thoughts, worries or rumination (cognitive symptoms) (Department of Education and Training Victoria, 2020).

- Maths anxiety may be present in the following ways in the classroom: students may experience working memory deficits, be slower to complete tasks, make more errors, and avoid maths.

Have you noticed your students experiencing maths anxiety in your classroom?

Take the time to think about the students you have taught. When have you witnessed maths anxiety? How did it present? Were there particular tasks that provoked a feeling of anxiety for some students in your mathematics lesson?

PART 4

HOW CAN I MAKE A DIFFERENCE IN MY OWN CLASSROOM?

6
Creativity in Mathematics

Creativity is a fundamental 21st-century skill (Bicer et al., 2020 & 2021; Novita & Putra, 2016). As outlined in Chapter 2, we live in an increasingly automated world where machines and robots are being used to replace humans to complete tasks in both the home and the workplace. Although the invention of household items like the robotic vacuum cleaner has saved many households significant cleaning time, the threat of automation in workplaces threatens jobs and livelihoods. Machines and robots cannot be creative (at this present time) because they are limited to performing the function they were programmed to do.

As humans, we are able to think and act in ways that are flexible, we are able to adapt to circumstances, and we are all unique – so unique, in fact, that no two people have identical fingerprints. We are able to have differing opinions and thoughts. We can see the same problem from different angles, and we can use different strategies to solve the same problem. Computers cannot do this. Robots and machines cannot do this.

The distinct advantage of humans is their ability to think creatively (Bicer et al., 2021). Contrary to popular belief that creativity is something miraculous or perhaps accidentally bestowed upon a chosen few, it is a uniquely human trait. Creativity is often associated with artistic pursuits, and some view it as a non-essential skill; however, creativity is a fundamental human trait that has allowed us to problem-solve, adapt, change, and innovate in many areas far beyond the artistic, throughout history.

Built over 4000 years ago, the Egyptian pyramids are a display of creative human ingenuity. The Ancient Egyptians created novel complex structures that required the application of sophisticated mathematics, science, engineering, and architecture. It took decades, and teams of thousands, to build these massive structures. With the sheer weight and size of the limestone used to build the pyramids, I am sure there were many mistakes along the way where the Ancient Egyptians were required to adapt, change, and use creativity to solve the problems faced.

Creativity has allowed humans to continue to innovate. We have designed various modes of transport, and in recent times we have used creativity to make transportation like cars more environmentally friendly and sustainable. It is essential that we promote the use of creativity in our society, classrooms, and workplaces now and into the future.

General workplaces are not traditionally viewed as places that value creativity. Interestingly, a survey of major CEOs suggested that 'creativity is the most important characteristic for success' (Kozlowski & Chamberlin, 2021, p. 97). Creativity can be used to view a problem from another angle, to innovate on a product that needs to be improved, or to design something that has never been invented. *The New Work Smarts Report* (Foundation for Young Australians, 2017) predicts that by 2030, further technological and automation advancements will increase the need for workers to think creatively.

In our classrooms, finding the time to allow students to think creatively can feel like a challenge. In the last section of this chapter, I will suggest ways that you can continue to satisfy the requirements of the Australian Curriculum while enhancing the opportunity for students to use creativity in mathematics. Creativity is a skill that can be developed when optimal conditions are strategically set up for students to explore problems, discuss ideas, and think of possible solutions in a supportive classroom environment that celebrates persistence and where joy is found in the struggle of challenging problems (Australian Curriculum, 2022; Boaler et al., 2020; Chamberlin & Mann, 2021; Novita & Putra, 2016).

Creativity in Australian education

Creativity is embedded in the Australian Curriculum across all learning areas as a General Capability. General Capabilities are one of the three dimensions of the Australian Curriculum that encompass the knowledge, behaviours, dispositions, and skills that aim to equip students in the classrooms of today for their futures. The General Capabilities of the Australian Curriculum include:

- Critical and creative thinking
- Digital literacy
- Ethical understanding
- Intercultural understanding
- Literacy
- Numeracy
- Personal and social capability.

Australian Curriculum (2022) reinforces the importance of the General Capability: Critical and Creative Thinking to give students opportunities to generate and apply ideas, to see situations in new ways, to identify alternative possibilities, and to create and link ideas.

I was fortunate enough to be part of the Queensland Curriculum Advisory Group from 2020 to 2021 in the review of the Australian Curriculum. I worked with a group of teachers to analyse, review, and provide targeted feedback on various elements of the curriculum review for our nation's students. Throughout this process, it was richly rewarding to learn more about the intent of the curriculum from its inception. It was interesting for me to learn that the Australian Curriculum was built on an ideological foundation of providing a national, world-class curriculum with the intent of equipping learners to become successful, confident, and creative individuals who will in turn use their learning in our schools to become active and informed citizens. Right from the inception of the Australian Curriculum, there was an intent to embed and value creativity.

The Alice Springs (Mpartnwe) Declaration (Australian Government, Department of Education, 2019) outlines a national vision for Australian education. The Declaration emphasises the need to improve outcomes for all Australian students, from every background and community across our vast nation. The Declaration acknowledges that we need our education system to meet the needs of the learners in our nation in a rapidly changing world. One of the two overarching goals of the Alice Springs Declaration is:

> 'All young Australians become confident and creative individuals, successful lifelong learners, and active and informed members of the community.' (Australian Government, Department of Education, 2019, p. 6)

Creativity is a central and valued element in education in Australia.

Take a moment to reflect

- How often do you see creativity embedded in practice throughout the curriculum?
- Do you think creativity is valued in our schools?
- Do you think creativity is valuable in mathematics?
- In what ways do you promote creativity in your mathematics lessons?

In 2022, the Programme for International Student Assessment (PISA) assessed 15-year-old students across the world on their creative thinking abilities (OECD, 2024). Students were tested on how well they created and generated diverse ideas. The test analysed how well students evaluated and improved upon other ideas in order to reach creative outcomes. Australian students scored extremely well and were shown to have strong skills in creative thinking, with Australia sitting fourth out of 64 countries in overall performance, and 88% of our students attaining the baseline proficiency in creative thinking.

Creativity in mathematics

The International Mathematical Union recognises two to four people every four years by awarding the Fields Medal. This prestigious award recognises 'sustained creative effort' (Mehta et al., 2016, p. 14) in mathematics. Maryam Mirzakhani (the first female Fields Medal recipient, awarded in 2014) epitomises creativity in mathematics. Mirzakhani's area of particular interest was geometry, where she preferred to work slowly, visualising problems and drawing on large sheets of paper on the floor while creating elaborate stories to work out complicated problems.

When Maryam's daughter was in kindergarten, she was asked what her mother did for work. Her daughter explained that her mother was an artist. Maryam's daughter saw her mother's elaborate mathematical drawings and firmly believed that her mother was an artist. Maryam used drawings to help her to make sense of hyperbolic geometry, and what Maryam uncovered by looking at complex problems with a creative lens earned her the highest honour in mathematics in the world.

Jo Boaler is an excellent example of someone who has combined creativity and mathematics to educate students. She has done extensive work in mathematics education across the years of schooling and in universities across the world. In her book *The Elephant in the Classroom*, Boaler (2016) reinforces the need to encourage students to work creatively and visually: 'It is very important to engage students in thinking visually about mathematics, as this gives access to understanding and to the use of different brain pathways' (p. 185). She goes further to say that visual representations and creativity in mathematics both support students therapeutically and in allowing access to tasks for all students. Jo has also developed a series of books for teachers to use in the mathematics classroom that are excellent examples of combining creativity and mathematics to transform student learning.

Cultivating creativity in maths lessons

Teacher as facilitator

Real mathematicians do not solve challenging problems with quick ease. Maryam Mirzakhani spent years trying to uncover the mathematical explanation for hyperbolic surfaces (from the time she spent during her PhD on this topic, which she completed in 2004, until her death in 2017). She needed time to test her theories and refine them, time to make mistakes and learn from them, time to discuss her thinking and apply it. Even after working on this specific area of mathematics for many years, Maryam had not completed her theory when she passed away.

There is joy in the struggle of being challenged. As teachers, we need to create classroom environments where students trust us enough to know that it is ok to make mistakes, test theories, discuss possibilities, and ask questions. In order to create this classroom environment, teachers need to shift their pedagogy. Instead of positioning themselves as the bearers of all knowledge, pouring information into empty student vessels to help them to learn, teachers need to shift their position to being facilitators of learning.

As a facilitator of learning, you must be clear about the intention of your lesson. Ultimately, you still need to guide the students' thinking and interactions. A facilitator is clear on what the purpose of the learning is, how it links to the curriculum, and what its goals look like. A facilitator guides students, asks carefully considered questions that enhance and encourage their learning, plans enabling and extending prompts to help the lesson flow well and cater to diverse learning needs, and creates an environment where students can discuss their thinking and try out their theories.

By standing back and refraining from dominating the lesson, we allow students the opportunity to think about their learning. I would like to stress that teachers do not act as facilitators if they rock up to the lesson without careful planning. Refining the art of teaching takes time in planning, adjusting, reflecting, and refining. It may feel uncomfortable or strange at first to take a step back from being the centre of the lesson, and there is still a place for quality, explicit instruction. I would, however, like to encourage you to start taking small steps to position yourself a little further out from the centre of the lesson each time. Could you ask a more open-ended question in your maths lesson that encourages dialogue and discussion? Perhaps you could practise creating enabling and extending prompts in your usual lesson and see how your students respond? Whatever you choose to do, it is powerful to witness students feeling more empowered to take control of their learning, and by taking even small steps you will allow more space for moments of creativity in your maths lessons.

Open-ended questions

Stop and reflect for a moment. Which question provokes a deeper response from you?

- Question 1: How many fingers are on your right hand?
- Question 2: How many different ways can you represent the number 5?

We know that there is a place in our classrooms for both closed questions and open questions, but often we can get stuck in the trap of choosing more closed questions that elicit quicker responses from students. Closed questions are great to help us to recall facts and single answers, but when we consider what the research tells us about the future that our students face over their careers, we can see that there is a great need to engage them in deeper thinking.

Open questions require higher-order thinking. They are likely to start with words like 'how', 'which' and 'why', with a varied range of possible responses. I acknowledge that closed questions can be a great way to recap key learnings in a lesson, or reinforce key facts that need to be reinforced, but I challenge you to keep a silent tally in your next maths lesson (or ask a colleague to keep one for you if you are in a co-teaching space) of how many of each type of question you ask in a typical lesson.

It can be really helpful to pre-plan your questions and keep these questions in a space near where you are teaching, to prompt you if you find yourself falling into the habit of asking a string of closed questions. If you find yourself sticking to closed questions, branch out. After a student responds, try to encourage deeper thinking by asking some of the following questions:

- 'Can you explain your thinking to me?'
- 'What makes you think that?'
- 'Can you tell me more about your answer?'

- 'Did anyone solve the problem in a different way?'
- 'How do you know?'

Let's try out the example closed question again in a classroom scenario:

Teacher: *How many fingers are on your right hand?*

Student: *Five.*

Teacher: *How do you know?*

Student: *I counted each finger. See (points to each finger and counts): one… two… three… four… five. I have five fingers.*

Teacher: *Who agrees? Did anyone count their fingers in a different way?*

Now, I know this example is one of simple mathematics, but even our youngest students should be encouraged to think about their responses. Yes, we want to encourage subitising and fluent counting, but we equally need to encourage students to critically reflect on how they reached their answer. The more comfortable students are with discussing their thinking and reflecting on how they solved problems in primary school, the better they will be at justifying and showing their working out in high school. As someone who has taught students from Prep to Year 9, I can tell you that one of the biggest confidence killers in high school is where students are penalised for not effectively communicating their thinking and reasoning in mathematics. Let's empower our students to think deeply about how they reached a solution and facilitate this for our students by asking questions that open up the learning and classroom discussion.

Rich tasks and maths investigations

As you are learning throughout this book, there is a critical need for students to be able to apply their maths knowledge in a range of contexts using flexible and efficient strategies. The days of celebrating the fastest learners are behind us; we now need to develop student confidence to apply a range of strategies to solve problems in creative and flexible ways. The need to encourage discussion is critical in developing a solid understanding that we can approach problems from a range of angles (and understanding that people may see the same problem in different ways) using a range of strategies and tools. Automation has taken away the need to recite numbers quickly off the top of our heads (although I would argue this was never a quality 'skill' that we encouraged in schools), and students now need to learn how to persist in challenging situations.

Since the Industrial Revolution, mathematics education has involved a passive acquisition of knowledge. Contemporary learners of mathematics require engaging and rich tasks for deep mathematical understanding. The Australian Association of Mathematics Teachers (2009) argues that school mathematics in the 21st century needs to contribute to students' development by providing knowledge and conceptual understanding through deep learning; develop students' ability and capacity to think and work mathematically in the real world; and be delivered in a context that makes meaningful sense to them. Henry (2015) reinforces the need for good pedagogy to 'build on what children already know and understand' (p. 2). Furthermore, Henry (2015) and Boaler (2022) argue that to participate in the digital age, children need to be given experiences that develop their problem-solving skills, critical thinking skills, and creativity, and that it is essential for 21st-century learners to have opportunities to collaborate and discuss their thinking with others. Boaler (2022) states that schools need to develop a culture that 'values [students'] ability to think and reason, invites their engagement in the breadth of mathematics, and encourages all students to achieve at high levels, even those who struggle for some of the time' (p. 34).

Rich tasks, therefore, can be summarised as encompassing elements of (Figure 8):

- Knowledge through deep learning (Australian Association of Mathematics Teachers, 2009)
- Working mathematically in real-world contexts, connected to what students know and understand (Australian Association of Mathematics Teachers, 2009; Henry, 2015)
- Problem-solving, creativity, and collaboration (Henry, 2015; Boaler, 2022)
- Inviting engagement in the breadth of mathematics (Boaler, 2022).

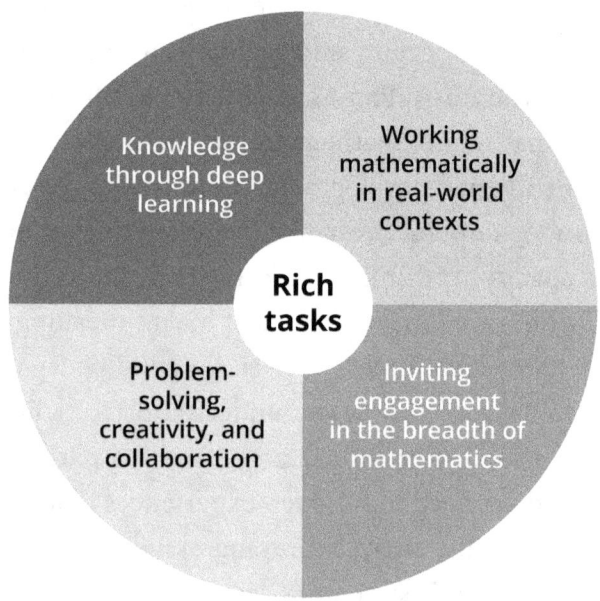

Figure 8. Elements of a rich task. This model combines four elements of a rich task from Australian Association of Mathematics Teachers (2009), Henry (2015), and Boaler (2022).

Furthermore, for students to be able to engage in deep learning, it is essential that teachers not only have good content knowledge, but that they can also get this across to students through an application of pedagogical content knowledge (PCK), to ensure that students are able to both learn and understand (Hill et al., 2008). Rich tasks themselves will not provide the depth needed for both learning and understanding if teachers do not have knowledge of what may make a topic easier or more difficult to learn and understand and ways to address this by using a range of alternative forms of representation, based on both research and the wisdom of practice (Shulman, 1986).

The selection of rich tasks should be 'low-floor, high-ceiling'. This means that the task allows every student to experience a level of success, with the potential for students to extend themselves built into the task. NRICH (2019) says that low-floor, high-ceiling tasks allow everyone to demonstrate what they can do, without worrying about what they cannot do. It is important to find the right type of task to fit your learners and context. It is very powerful for students when everyone is given the same task and no-one's achievement is limited before even beginning. Boaler (2016) argues that 'students who struggle in Mathematics are helped when they engage in discussions about maths with students who are working at higher levels' (p. 104). Rich tasks will provoke curiosity, spark discussion, and promote student engagement in creative ways. Tasks are selected to build on what students already know and understand and provide a level of challenge.

Rich task example: the Foot Parade

Being able to make 10 is an essential foundational skill for students. Unifix cubes, tens frames, and counters are widely used by teachers to show how 10 can be decomposed in a range of ways. While these hands-on tools are useful, this type of task is usually teacher-led and explicit. These tasks do not allow for deeper levels of engagement, discussion, and a development of fluency and flexibility with numbers.

Through critical reflection, teachers can transform a making 10 activity into a rich task that ensures that students gain knowledge through deep learning; work mathematically in real-world contexts; connect to what they know and understand; use problem-solving, creativity, and collaboration; and engage in the breadth of mathematics. An example of task transformation to make 10 is shown in Figure 9 – the 'Foot Parade'. This task invites students to engage not only in a breadth of mathematics but also in a range of General Capabilities.

Figure 9. Foot Parade task (adapted from Boaler et al., 2020).

The Foot Parade is a rich task that engages students in elements of the Australian Curriculum: General Capabilities for Critical and Creative Thinking and Mathematics (Australian Curriculum, 2022). Students generate and apply ideas to translate their knowledge of making 10 using a range of number combinations; they look for patterns, identify components, and organise animals. Through exploration and experimentation, students see the composition of the number 10 in new ways. They predict and draw conclusions by rearranging animals in groups; substituting known animals for unknown animals; discovering what other animals have four legs and if any animals have 10 legs; and discovering the many different possible combinations for the Foot Parade.

Foot Parade: A Creative Mathematics Task for Foundation	
Possible Mathematics Content Descriptions	
AC9MFN01 – name, represent and order numbers including zero to at least 20, using physical and virtual materials and numerals	
AC9MFN03 – quantify and compare collections to at least 20 using counting and explain or demonstrate reasoning	
Critical and Creative General Capability	Personal and Social General Capability
• Develop questions • Identify, process and evaluate information • Create possibilities • Consider alternatives • Put ideas into action • Interpret concepts and problems • Draw conclusions and provide reasons • Evaluate actions and outcomes • Think about thinking (metacognition) • Transfer knowledge	• Personal awareness • Reflective practice • Perseverance and adaptability • Communication • Collaboration • Leadership • Decision-making • Conflict resolution

Figure 10. An analysis of 'Foot Parade' against the Australian Curriculum.

We have been implementing the Foot Parade task in our school for a couple of years now, and each time our Prep teachers conduct this lesson, they are so excited to learn more about their students. The students are always deeply engaged and excited, and this task beautifully exemplifies the balance of all of the elements of a rich task. Before delivering this lesson, our teachers have clear goals and intentions for the task (aligned with the curriculum). As I recommended earlier, it is good practice to pre-construct questions that will both enable students to achieve success and extend their thinking – questions that will help to guide the lesson – and to have some key questions that you would like to ask to uncover any misconceptions.

Some of the questions I would suggest you consider asking your students as they explore the solutions to this task include:

- What is the longest parade you can make with 10 feet?
- What is the shortest parade you can make with 10 feet?
- Design a foot parade for your favourite animal to fit in.

By using the Foot Parade as a sample rich task, the Prep teachers in my school were amazed at how deeply students engaged in the task. Students showed persistence in trying to find solutions, they demonstrated resilience and deep understanding of decomposition, and they creatively extended their thinking by finding a variety of solutions throughout each lesson.

In the first lesson, students worked hard to count every single leg of the animals to make a Foot Parade (see Figure 11). Students were excited to show their Foot Parade to their peers, and the teacher took digital images of each parade to show the class on a smart panel. The teacher scaffolded the key language and concepts of addition and labelled student work to support student understanding of additive thinking.

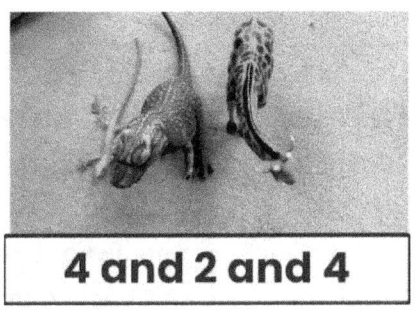

Figure 11. A student photo of a Foot Parade with a lizard (4), dinosaur (2), and giraffe (4).

For a task to encourage creativity, there needs to be depth and scope in the task that allow for students to respond in a variety of ways. In the second lesson, students were keen to explore how many different types of Foot Parades they could make. They worked in pairs to make a range of different Foot Parades (pictured in Figure 12) and they were excited to share their discovery about the different combinations of legs and animals that made a Foot Parade. The teacher then used mathematical language to model the labelling of each of the parts of the parade and discussed how each of these parts combined to make 10 altogether.

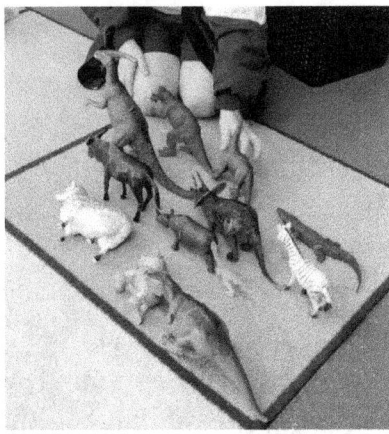

Figure 12. A student photo of four different Foot Parades.

In the final lesson, students explored many different types of animals and sorted them in a class discussion according to how many legs each animal had. Students then used movable images in an online learning platform to create and label more animal parades. In the final lesson, the teacher modelled key additive concepts and demonstrated how to label each part of the Foot Parade (as shown in Figure 13, where a student has labelled each animal in their parade with a number and then calculated the total to 10).

Figure 13. A screenshot of a student's digital Foot Parade.

The Australian Curriculum has been written to elicit more depth of learning than the surface level of teaching a Content Description. By using a task like the Foot Parade, teachers are not only teaching essential skills and content from the mathematics curriculum, but through the pedagogical decisions made to deliver this content in this way, problem-solving and reasoning are also developed. Students are given opportunities to think creatively as they explore possibilities, consider alternatives, and put ideas into action.

This example of a rich task spans a few lessons. A rich task could be like this example, spanning more than one lesson, or it could be a carefully created question to lead students to investigate a mathematical problem in 10 minutes. The size of the task and the time the task takes are not what is most important; it is the richness and depth of the task that matter most. I challenge you to look at one of the lessons you plan to deliver in your classroom this week. Are there ways that you can increase creativity in your lesson? Could you tweak an activity you have planned to make it richer? Are there some questions that you could write to open up discussion in your classroom?

CHAPTER SUMMARY

- Creativity is a fundamental 21st-century skill that has allowed humans to innovate throughout history and will ensure increased job security for our students in the future.

- Creativity is embedded within the Australian Curriculum and is a valued element in Australian education.

- 2022 PISA data showed that Australian students are strong in their creative thinking abilities, with 88% of the 15-year-olds tested attaining the baseline proficiency in creative thinking.

- We need to create classroom environments where students feel safe enough to make mistakes.

- Teachers need to position themselves as facilitators of learning, being clear about the intentions of each lesson, scaffolding carefully created questions that guide students in their learning and elicit discussion.

- Rich tasks and maths investigations provoke curiosity, spark discussion, and promote student engagement in creative ways.

- Rich tasks can be as small as a carefully crafted question that students explore over 10 minutes or as big as a task that spans more than one lesson.

7

Curiosity in Mathematics

As a school leader, one of the most joyous parts of my role is welcoming new students and their families to our school community. I have the privilege of conducting enrolment interviews, meeting the new students in Prep Orientation Week, and presenting information to families at our Parent Information Night. One of the most common concerns for parents is that their child may not yet know how to write their name, count to ten, or say letters of the alphabet. One of the things I love about our school is that we don't consider these skills to be necessary to learn before children begin Prep (although writing their name can be very handy in order to label things that are important to them and ensure that their work is not mistaken for someone else's). I assure parents that their child will learn these things at school and that it is extremely valuable for the family to create a love of learning and encourage curiosity. Nurturing and encouraging a child's natural curiosity will hold them in good stead throughout their schooling and into adulthood.

I love being able to chat with these small children and hear about the things they like to do, the stories they like to read, and their favourite things. It is a magical moment when you find the topic that really lights a child up. You see the look on their face as they share their passions with such enthusiasm and wonder. Often, I will then ask the child further questions to see what they know about their special interest, and encourage them to learn more about it. They can barely wait to finish the interview so that they can get home and find out more! The spark of curiosity is truly a precious thing to behold.

In our classrooms, we can often feel burdened with the content required to be taught across all learning areas and dimensions of the curriculum. It can be easy to fall into the habit of teaching things using a checklist to ensure that each area is covered, and it can feel hard to find time to stop and wonder, ask questions, and explore topics further. Building a culture of curiosity begins with asking great questions and encouraging dialogue, and I would argue that this curiosity should be encouraged beyond just scheduled lesson time. I have seen some wonderful examples of teachers displaying students' questions at the beginning of a unit of learning, but I think it is just as valuable to have a piece of butcher's paper on the wall that students can continue to add to using pens, highlighters, and Post-it notes as they learn new concepts. I call this type of collaborative writing/drawing space a 'Wondering Wall' (see Figure 14). It is a place where aesthetics are not important, but where student agency is valued as students are encouraged to contribute thoughts, questions, and when uncovered, answers to those questions.

> **Wondering Wall**
>
> A collaborative writing and drawing space for students to record their mathematical thinking and wondering.
>
> Suggested resources to make a 'Wondering Wall':
> - a piece of butcher's paper (secured to a wall in the classroom)
> - pens
> - highlighters
> - Post-it notes.

Figure 14. A summary of a class 'Wondering Wall'.

More than any skill or level of academic knowledge, I believe that curiosity is one of the most important things for us to value and help to grow in our students. Curiosity helps us to ask deeper questions, it motivates us to find the answers, it fuels the desire to deepen our understanding. By encouraging curiosity and building a curious culture in our classrooms, we are encouraging our students to actively engage with learning, to share their ideas, seek out new knowledge, and connect what they learn to their lives. Curious classroom cultures encourage persistence through challenging topics and concepts, and create an environment where students don't engage in learning because they have to but because they want to.

When a culture of curiosity is built, curious learners do not simply settle for surface-level answers. Curious learners challenge assumptions, consider different perspectives, and over time develop the skills to analyse and synthesise information. Classroom cultures of curiosity encourage students to take ownership of their learning, often going beyond what's required in the curriculum because they're invested in the topic.

Building a culture of mathematical inquiry

Curiosity brings a sense of wonder, joy, and excitement to learning. It transforms learning from mere instruction to an exploration. However, this process of encouraging exploration and questioning of concepts can only be successfully established in a classroom when the teacher ensures that students feel safe enough to take risks in their learning and, most importantly, when they feel safe to make mistakes. This type of learning environment is set up through the lens of what Standford University professor and well-known author Carol Dweck calls 'Growth Mindset'.

Dweck (2017) argues that there are two ways that students can view their intellectual abilities, otherwise known as two mindsets. Students with a *fixed* mindset believe that their abilities are fixed. These individuals desire to look 'smart' and will avoid challenges, give up easily, avoid making mistakes, and not be willing to see the perspectives of others in their pursuit to get things 'right'. People with a *growth* mindset, on the other hand, believe that they can learn from the perspectives of others, embrace challenges, and view mistakes as opportunities to learn and grow.

When building a culture of curiosity and mathematical inquiry, it is critical to first build a classroom culture that values a growth mindset. Curiosity is a journey of discovery and is not a quick and simple question-and-answer process. Encouraging a growth mindset requires a safe classroom culture where students know that it is ok to make mistakes and take risks. You, as the teacher, must also model this growth mindset and model making mistakes and learning from them, finding out answers to questions, and contributing to wonderings. Your role in the creation of the culture is vital, and the students will look to you to see how to cope when they face challenging situations in their learning.

Maths investigations are wonderful for building this classroom culture of curiosity, practising persistence in the face of a challenge, and making connections to real life. Great maths investigations include the elements of a rich task (see Chapter 6); are low-floor, high-ceiling to set up all students with a level of access to the task and ensure success for all of them; are supported through teacher questioning, using both enabling and extending prompts; and engage students in higher-order thinking skills.

There are many wonderful examples of maths investigations that are easy to find, but to ensure the right selection of these tasks, teachers should critically analyse the resource to check:

- Does the task align with the curriculum?
- Does the task align with my learning intentions?
- Does the task have a 'low floor' entry point, to allow all students to have access to the task, and does it allow for a 'high ceiling' so that students can extend their thinking?
- If the task does not have the above, what do I need to change to ensure that all of the learners in my classroom have equitable access to the task?
- Is this task rigorous enough to engage students in higher-order thinking?

Bloom's Taxonomy (see Figure 15) is a hierarchy of skills that reflects growing complexity and ability to use higher-order thinking skills. It is a useful framework to support teachers in reflecting on the depth of the tasks, lessons, and units that they are planning to ensure that learning goes beyond memorising facts and promotes deeper understanding. In the previous chapter, I discussed a task that my Prep teachers implemented in their classroom named 'Foot Parade'.

The Foot Parade task engages students in various levels of thinking in accordance with Bloom's Taxonomy. In this task, students move from knowledge and facts (where they have learnt how to count to 10 and make 10) and comprehension (where they translate their knowledge of making 10 using a range of number combinations) to the more complex skills of analysis (where they look for patterns, identify components, and organise animals) and synthesis (where they create new ideas of how to make 10 and where they predict and draw conclusions by rearranging animals in groups, substituting known animals for unknown animals, discover what other animals have four legs and if any animals have 10 legs, and discover the many different possible combinations for the foot parade).

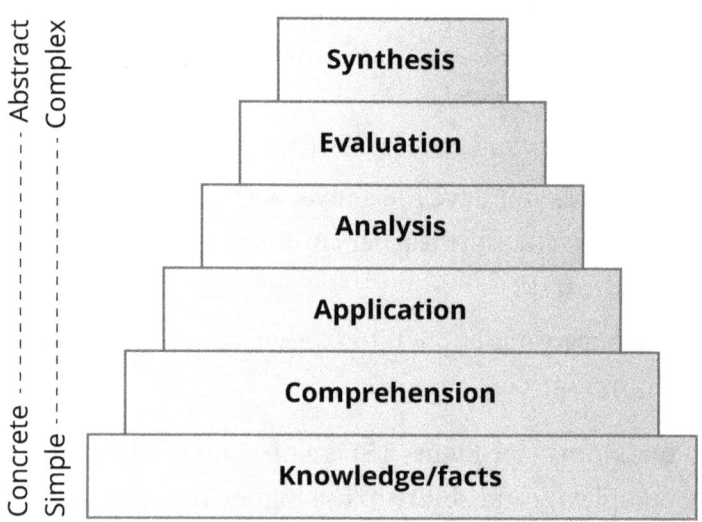

Figure 15. This figure was produced by Piggot (2017) from B. S. Bloom (Ed.). (1956). *Taxonomy of educational objectives: The classification of educational goals*. Longmans, Green.

As discussed earlier in this book, and as illustrated in Figure 15, it is important to go beyond surface-level thinking with students, where they learn facts and knowledge, and provide them with experiences that allow them to engage with challenging content. Maths investigations should go beyond simply solving a problem and encourage students to engage in both thinking and reasoning. Reasoning, or explaining why, is sometimes a forgotten component in primary classrooms, but is essential to ensure our students deeply understand and can apply their mathematical knowledge beyond familiar contexts and generalise this knowledge in unfamiliar and more complex contexts as well.

Making connections to real life

Making learning contextual, connected, and relevant to our students is important. Students can engage more deeply in an investigation or task when they feel connected to the learning. They need to feel like there is a reason for putting in the effort and that this learning may benefit them in their real lives.

Firstly, students need to feel like mathematics is part of everyday life. We need to keep our eyes open for examples of how mathematics was used by people in the past to innovate (e.g., creating and using sundials to tell the time), is being used to solve current problems (e.g., using mathematical modelling in forensic science), and may be used in the future to benefit our society. Students need to see that mathematics is an integral part of daily life and is critical to our world and society. The more examples we can give students to show them how mathematics has helped to shape our world and the potential for maths to help our world in the years to come, the more they will understand the importance of developing their own mathematical knowledge.

Teachers need to be mindful about selecting maths investigations that reflect the context and backgrounds of their students. If your students live remotely, connect the context to what they know and experience in their day-to-day lives so that it feels like the learning is purposeful and meaningful to them. If your students live in inner-city apartments, select tasks that reflect their context to allow for meaningful connections. A big note here is to make the extra effort to connect learning to the diverse backgrounds of our students. We have students with a range of different cultural backgrounds at our school, and we make every effort to learn more about these cultures and try to respectfully reflect these cultures in the tasks, resources, and experiences that we give our students.

Students need to be able to see themselves in the learning. To foster deeper engagement and meaningful learning, it's essential to make mathematics contextual, connected, and relevant to students' lives. When students see a purpose in what they're learning – especially how it relates to real-world situations – they are more likely to be motivated to actively participate in lessons and persist in challenging tasks. Selecting mathematical investigations that reflect students' lived experiences and cultural backgrounds makes learning more meaningful and purposeful.

Provocations to stimulate thinking

In order to build a culture of curiosity, we, as teachers, need to set up situations that stimulate students to want to learn more. A great way to establish the right environment for this is to use provocations to stimulate thinking and discussion. When I think of a provocation, I actually link it to my favourite early childhood pedagogical philosophy drawn from Reggio Emilia, called 'Invitations to Play'. In an invitation to play, teachers create settings to encourage students to engage in play and to interact with the materials in the environment.

I loved using invitations to play when my own children were little at home. The most basic way to explain an invitation to play is: instead of leaving a box of blocks out on a mat, you leave the box of blocks out on the mat and also leave some builder's hats and tools with the blocks, or set up some blocks as the start of a road with some cars there. Invitations to play guide children, and invite them to play and create in what I can best describe as the start of a story or scenario that they can then choose to play in. As a parent, these invitations ensured that the toys my children had were put to good use and didn't get boring for them, and helped to facilitate play and interaction between my children.

A provocation is similar to an invitation to play, but it is applicable across ages and year levels, whereas an invitation to play is aimed at younger children as they learn to navigate and explore the world through play. A provocation creates an optimal environment that literally provokes students to think and perhaps feel something. Provocations can be used in any subject and, when chosen carefully, can be a great launching pad to begin units of work, to stimulate discussion, and to frame lessons. Some people may also use the word 'hook' to describe a provocation when you select a resource, question, text, image, etc. that hooks students in to the learning.

Provocations surround us every day. Provocations are the things that make us wonder, think, and question. Recently, South East Queensland (where I live) made preparations for a tropical cyclone to directly cross the region for the first time since 1954. South East Queenslanders are no strangers to flooding events and storms, but for the first time in over 70 years, we were forced to prepare for this severe weather system to pass through our towns and cities. I asked many questions as I grappled with the food we needed to buy, the way we needed to tape our windows, the preparations we needed to make for a safe room in our house should the cyclone become more intense, and how we could shelter in this small space. I was filled with so many questions, as was the community around me. I was thinking, questioning, and wondering, and I learnt so much during these preparations. In the end, the cyclone was downgraded to a tropical low, but the learning, information sharing, and doing along this journey of cyclone preparation will be with me for life.

I am certainly not suggesting that you organise a natural disaster to get students in the zone for learning! Rather, think carefully about how you could provoke thinking, questioning, and wondering. It is important to have clear learning intentions as you select provocations that are authentic to the students.

Trevor Mackenzie, teacher and author, believes that there is great power in provocations. He says that provocations are so powerful that they should stir thought and wonder in our students and create optimal conditions for them to engage in learning with questions and curiosity. Mackenzie (2018) goes further to state that provocations:

- Illuminate questions
- Radiate passions
- Intensify investigations
- Heat up prior knowledge
- Spark curiosities
- Enliven engagement
- Brighten interests, and
- Switch on wonders.

I agree that provocations provide the many layers of richness in learning that Mackenzie outlines above. There are many ways to provoke student thinking and hook them into their learning in mathematics, but I will focus on an area that I think best illustrates the need for students to sharpen their knowledge and understanding, and that is interpreting data.

In 2018, it was found that '90% of the world's data was created in the last 2 years' (Marr, 2018). Critical analysis and investigation of data is essential to 21st-century learning. Henry (2015) argues that while 'students may come to school with confidence and competence in using technologies, they do not necessarily come with fully developed analytic and evaluative skills' (p. 6). The students in our classrooms need to develop critical skills to enable them to interpret, analyse, and evaluate data. Understanding data and being able to critically analyse information have become even more important with the sharp rise in fake news and the use of artificial intelligence to generate information that appears real and credible but is often inaccurate or entirely false. One of the most effective ways to teach students to critically analyse data is through Data Talks.

The best examples of Data Talks that I have found are on Standford University's *You Cubed* website, where a range of excellent, provoking data sets have been carefully selected to stimulate questions and discussion. My favourite Data Talk on this website is called 'Hand Washing'. This visual data representation from a study of health professionals in 1978 shows the effectiveness of the health-care workers' hand-washing techniques. The visual shows the back and front of a hand and the different areas of the hand that are most often, often, and less often missed in hand washing. The free download has some great guiding questions to help you facilitate a Data Talk. I think that in our post-Covid world, we are much more conscious and careful about hand hygiene, and this Data Talk can stimulate great discussion. If the same study was conducted today, what would the results be? And again, if the study had been conducted in 2020, what would the results have been? How would each of these data sets compare to each other, and what could this data tell us about hand washing?

I enjoy curating and collecting a range of interesting provocations. Head to my Instagram page, @Primary_Pedagogy, where I continue to add to a collection of thought-provoking and stimulating resources for the classroom.

Engaging students using curiosity in mathematics

A few years ago, I was sitting with a group of teachers at my school who were venting their frustrations with trying to build depth and challenge into maths lessons using informal measurements. The class had experienced a year with a number of changes of teachers, which had meant that in the final term of the year there were lots of 'bits' of the maths curriculum left to teach and assess. As any teacher knows, the fourth term is the final reporting term of the year, which can bring a certain pressure to finalising coverage of the curriculum, particularly ensuring coverage of the Achievement Standards. Along with this, Term 4 sees lots of exhaustion among both students and teachers, with end-of-year school events, swimming lessons, and the like adding to the feeling of there being insufficient time to teach what needs to be taught.

The group of teachers I spoke to had mapped out the term, covering a new concept each week, and had designed a portfolio of assessment pieces to complement the learning sequence. This was a good idea to ensure coverage of the content that needed to be taught, but the teachers were concerned that there was not enough depth built into the lessons and that they would move too quickly to ensure that students gained an understanding of the content. I agreed with them, and we decided to look at their scope and sequence of content to find a better solution.

By looking at what needed to be covered as a team, and by having a collaborative discussion, we ended up designing three smaller units of work that both covered the curriculum and linked key concepts in meaningful ways. We went a step further to ensure that each unit was connected to students' real lives, and that it uncovered misconceptions, so that the teachers could adjust their teaching to ensure that all students understood the concepts, and we brainstormed a range of open-ended questions to open up the learning and build a culture of curiosity.

The case of the missing bed

One of the concepts that the teachers did not want to rush through was teaching students about flips, slides, and turns. The team had taught Year 2 before and noted that their students often found this concept to be difficult to grasp and abstract. They reported that they had difficulty finding the best types of concrete materials to use to demonstrate this concept well to students and found that it was a concept that was seemingly simple to teach but actually tricky to teach well.

Another concept that had not yet been formally assessed during the year was informal units of measurement. The teachers had taught this in a range of ways throughout the year, but they were concerned about whether they had given the students enough opportunities to measure diverse objects using a variety of informal units. Did the students know how to use the most effective and efficient tools to measure informal units? Could the students explain their thinking and justify why they would choose one tool over another?

Finally, the students had not been assessed on their ability to interpret simple maps of familiar locations. At first, these concepts seem disconnected, and I can see how they were sequenced out over three weeks. It made logical sense to ensure coverage of the curriculum. But would teaching these three concepts in separate silos ensure deep, connected learning? Would the students not only know the concepts but understand them by the end of each week? Would the students be able to transfer their learning of these concepts across both familiar and unfamiliar contexts in their next year level, and indeed in the years to come? I am not convinced that teaching these concepts in quick succession would have given students the opportunity to deeply understand. Instead, we combined the three concepts into one rich unit of work spanning the three weeks allocated (time was not on our side, but no matter the timeframe we had, our learners were central to our decision-making). And so, 'The Case of the Missing Bed' was born.

On the day of the first lesson, the students entered the room to find a bed in the centre of the room. The bed had a blanket, a pillow at one end, and a teddy bear placed next to the pillow. Talk about building curiosity in the lesson! The teacher invited the students to sit in a circle around the bed and, using carefully crafted open-ended questions, guided a discussion about the bed.

- What is a bed doing in the middle of our classroom?
- Does anybody in this room own this bed? Who could this bed belong to?
- Can you describe what you see?

The teacher allowed the students to exercise curiosity and used some of their questions to guide parts of the discussion. And boy, did the students have lots of questions and lots of things that they were keen to discuss! It was important to discuss the bed in detail, including describing the end the pillow lay at, the side of the pillow the bear was on, and the sides of the bed that had the blanket carefully tucked under, because in the coming lessons the bed would reappear in new locations to demonstrate slides, turns… and flips! The students needed to be very familiar with the location of the bed in order to understand what changed and what stayed the same when the bed changed position in the room.

Next, the teacher prompted the students to look carefully at some of the informal measurement tools that she had placed near the bed.

She guided the students through a discussion about the different types of tools, and brainstormed about other tools (like hands and feet) they could use to measure the bed and its items. Students then worked in small groups to choose informal measurement tools. They worked together to use these tools to measure the pillow, the teddy bear, the blanket, and the mattress. The teacher gave the students time to discuss, experiment, and record their findings.

After the students measured each item, the teacher facilitated a discussion where students described the tools they had used to measure each item. She encouraged the students to compare their tools and prompted them to reflect on how much time it had taken them to use one tool in comparison to another (e.g., 'Did it take you longer to measure the pillow using Unifix cubes than hands? Why do you think this might be?') and how accurate one tool was in comparison to another. After collating the data and discussing the most efficient and accurate tools to use based on the size of the item, the students then used the data to predict who the bed belonged to and why.

The mattress was single and child-sized. This size of mattress was purposefully chosen for this activity because we wanted the children to connect to the item and draw conclusions based on their own egocentric worldview. I do not use the word egocentric in a derogatory way that implies that children are selfish; rather, we know that at younger ages, children have thinking that is very centred around themselves and their world, so in order to help them to make connections to their learning, we need to select activities that children see themselves in and can see a direct connection to. The students then used their individual maths journals to record their thinking and reflections (maths journals are discussed in more detail in Chapter 9).

Throughout the next several days, students would enter the room after a lunch break, after a specialist lesson, or at the beginning of the day to find the bed in a different part of the room. The bed was always set up in exactly the same way (with the pillow at the same end, the teddy bear on the same side, the blanket tucked into the same sides), but each time the bed demonstrated a slide, turn, or flip. Students became excited to find the bed and describe its position in relation to the first position it was in during their first lesson. As a result of these lessons, the students performed very well in their assessment, as they no longer saw slides, turns, and flips as abstract and unrelatable, but they deeply understood each transformation, as they had been encouraged to explore, question, and discover each of the movements of the bed.

Bed mapping

Seeing the learning through the eyes of our students in the planning phase of the unit, we knew the students would be keen to discuss their own bedrooms and beds. We wanted to build in meaningful connections to the students' real lives, and we designed a series of lessons around bedrooms. We focused on bedroom mapping, and we provided a range of activities for the students to draw their own bedroom using a list of key features. Building parameters around this task was important, as the intention of the learning was around mapping, the position of objects, and the bird's-eye view perspective of these shapes. If we did not set clear intentions for the learning, an exploration of this topic could lead to an hour-long debate over Bluey versus Disney Princess bedcovers, which is not conducive to productive learning. The parameters we chose to guide the lessons around bedrooms included:

- Using a circle to symbolise a bedside lamp
- Using a large rectangle to represent a bed

- Using a small rectangle to represent a pillow
- Using a small square to represent a cushion and a larger square to represent a bedside table.

Once the students had completed the set tasks using the parameters outlined above, they were then allowed to design a quilt cover and draw a small toy (neither of these features were assessed).

Throughout these lessons, students:

- Viewed bedrooms from a range of places around the world, discussing the features of each room, the furniture in each room, and the position of the bed in the room
- Discussed and explored a range of examples of bedroom maps (using the parameters outlined above)
- Used grid paper (with large squares) and cut-out shapes to move and place objects in rectangular 'bedrooms'
- Drew a map of their own bedroom
- Designed their own dream bedroom.

We're going on a bed hunt

The culminating activity for this unit of work, where students went on a 'bed hunt' around the school, combined the elements of informal measurement, shape, and locations on a map. Students received an A3 school map with plenty of room for them to draw on. They walked around the school with their maths journals, in small groups, with the support of an adult to get them safely from place to place. They followed a map, as a group, to navigate the school and find the hidden beds.

Around the school, a range of staff embraced the bed hunt. There was a large double sofa bed in the library, with a teacher asleep in it in her pyjamas. Members of school leadership brought in pyjamas, inflatable mattresses, cuddle toys, blankets, and sleeping masks to set up bed spaces in various places around the school. There was a round dog bed with a small dog toy in it in one room, a small doll's cot in another room, a child's mattress set up in another room, and a baby's port-a-cot set up in yet another room.

Students walked around the school following their map and discovering each bed. Most of the occupants chose to 'sleep' while the students explored each bed, and students quietly discussed what they observed (the shape of the bed, size of the bed, description of the bed, position of the bed) while they sketched their findings in their maths journal. Throughout the bed hunt, I observed students engaging in meaningful discussions, drawing labelled diagrams, using informal units of measurement to measure and compare the different beds, and carefully observing the placement of each bed to inform what they would later illustrate on their A3 maps.

Students were deeply engaged throughout this unit, and the activities chosen were transformative for their learning. They were very successful in their assessment tasks and demonstrated a deep level of understanding of each concept taught. The students loved having a personal connection to the content and being able to share about their own bedrooms. The unit of work was the start of a flame of curiosity that turned into a fire that burned more brightly than any of us at the original planning table could have foreseen.

Not only was this unit of work successful in terms of student engagement and learning, but it was also deeply rewarding for the teachers to be able to deliver lessons that they too were genuinely excited to teach. Think back to the beginning of this section of the chapter, where I described the three separate areas of content that had been mapped out to be covered over three separate weeks with a silo approach to learning. Can you see how these lessons were transformed into one cohesive, rich, and engaging unit of work? Reflect on your own lessons and context: is there one maths lesson that you could adapt to encourage the spark leading to a small flame of curiosity?

CHAPTER SUMMARY

- Curiosity helps us to ask deeper questions, it motivates us to find the answers, and it fuels the desire to deepen our understanding.

- Curious learners challenge assumptions, consider different perspectives, and develop the skills to analyse and synthesise information.

- Students with a fixed mindset believe that their abilities are fixed. Students with a growth mindset believe that they can learn from the perspectives of others, embrace challenges, and view mistakes as opportunities to learn and grow.

- Maths investigations go beyond simply solving a problem, as students engage in both thinking and reasoning.

- We need to give students examples of how mathematics has helped to shape our world throughout history and in recent times, and the potential for maths to help our world in the future.

- Students need to be able to see themselves in the learning. It is essential to make mathematics contextual, connected, and relevant to students' lives.

- Provocations stimulate thinking and discussion about the things that make us wonder.

8
Playful Maths

Play is arguably one of the most important elements of childhood development. Through play, children learn resilience, persistence, problem-solving, collaboration, and more. Play is where children role-play and make sense of the world around them and their experiences; it is where they practise their skills to refine them. Play is essential for cognitive, social, and emotional development, and countries like Finland – often top performers in international standardised assessments – attribute their success to the protection of childhood through the prioritisation and valuing of play.

Play is an evidence-based pedagogy, particularly prominent in early childhood education. Rooted in the work of Piaget, it supports how young children naturally explore, understand, and make sense of the world around them. In mathematics education, play offers far more than just enjoyment. Play fosters essential skills such as collaboration, communication, and metacognition. When students engage in playful learning, they are more likely to take risks, experiment with trial and error, make and learn from mistakes, and develop a sense of curiosity and joy in their learning.

Research has consistently highlighted the wide-ranging benefits of play in the early years. Specifically in mathematics, play provides meaningful opportunities for students to practise and apply concepts. These hands-on, authentic experiences help students to understand concepts (DePascale et al., 2023; Ramani & Eason, 2015). Ramani and Eason (2015) assert that the active, engaging, and student-centred nature of play can make it more effective than traditional didactic teaching approaches.

Metacognitive strategies are embedded in playful learning, which allows students to shift from being passive recipients of knowledge to active thinkers. Students begin to question, reflect on their own understanding, discuss their reasoning, and become more open to feedback. This dynamic form of learning invites teachers to respond in timely and direct ways to students' needs. Paatsch et al. (2024) reinforce that play promotes social interaction, collaborative problem-solving, and metacognitive development.

Although broad research into play-based mathematics pedagogy is still emerging, a study by Murtagh et al. (2022) provides compelling evidence of its impact. In a large-scale study involving 1691 primary students across 12 schools in Palestine, teachers who received targeted professional development and support in play-based learning saw measurable improvements in student academic achievement after just two terms. The study highlights the powerful role that professional learning and ongoing support can play in improving student performance, and the positive impact play has on student learning.

Parlakian (2022) emphasises the importance of foundational maths experiences from an early age. Drawing on a review of literature, she outlines five guiding principles for nurturing early maths through play: encouraging curiosity, embedding mathematical language throughout the day, mathematising everyday interactions, connecting learning to the home environment, and using storybooks to introduce concepts. While mastery of mathematical skills is not expected in the very early years, Parlakian argues that these rich early experiences can lay a strong foundation for future learning and help foster a lifelong enjoyment of mathematics.

How might play have a positive impact on student learning and engagement?

Play-based learning is recognised as an integral and valuable pedagogical approach in early childhood settings and is used in some primary schools. Research conducted in pre-school settings highlights the cognitive, emotional, and social benefits of learning through play (DePascale et al., 2023; Ramani & Eason, 2015; Murtagh et al., 2022). When students engage in playful, meaningful tasks, they are more likely to persist, take risks, and enjoy the learning process. This holistic engagement in learning contributes to deeper understanding and improved outcomes in mathematics.

Interest and engagement are key factors in student success. Doctoroff et al. (2016) found that children who demonstrated high levels of interest in mathematical activities tended to achieve higher mathematics scores. Playful approaches support positive and self-motivated engagement by making learning feel relevant and enjoyable. In addition to this, students who feel connected, supported, and safe are less likely to experience maths anxiety. Creating playful learning experiences in mathematics may serve as a powerful antidote to the fear and avoidance that can be associated with mathematics.

Modern mathematics education is moving away from passive, rote learning toward approaches that emphasise critical thinking, creativity, and collaboration. Boaler (2022) argues that students must be given opportunities to reason, problem-solve, and communicate their thinking in order to build deep mathematical understanding. She urges schools to develop a culture of mathematics that encourages active engagement with ideas.

Supporting this, Hattie and Zierer (2018) identified metacognitive strategies as one of the most impactful influences on student learning. When these strategies are embedded in classroom practice (particularly through the use of play as a pedagogy) students become more reflective, open to feedback, and capable of articulating their thinking. Play, as noted by Paatsch et al. (2024), naturally fosters these metacognitive behaviours, along with increased collaboration and social connection.

Teacher beliefs also play a critical role in how effectively play is used in mathematics learning. Anders and Rossbach (2015) found that teachers' past experiences and attitudes towards mathematics significantly influenced their ability to identify and support mathematical learning in play. This finding underscores the importance of professional development that addresses both pedagogical content knowledge and teacher confidence.

Play as a pedagogy is relevant to the teaching and learning of mathematics in our primary school classrooms, especially in the early years of primary school. Play is not simply an 'add-on' to traditional instruction; rather, it is a multi-dimensional and rich pedagogy that can transform mathematics learning. By embedding playful learning, we can build classrooms where all students are engaged, challenged, and joyful in their mathematical learning journey.

Case study

Before I started my master's research project, I was sitting in a webinar and heard a statement that early mathematical ability has been found to surpass early reading skills as a predictor of later academic achievement (Clements et al., 2023). I have been a teacher for almost two decades, I am a life-long learner, and I am always reading, attending professional development, listening to podcasts, and growing my knowledge using credible, well-informed sources of information. How had I never heard this before? How had the wider teaching population not heard this before?

These pieces of research (along with another 47 cited sources) became the driving force for the project that I conducted for my master's. However, I found it very difficult to find research in early primary school years, particularly in Prep. With the clear critical need to develop mathematics skills in strong and intentional ways, I was curious as to why there was so much research in early childhood contexts but scant research in primary school contexts, especially given that primary schools have these students from around the age of 4.5 until seven, where research clearly showed a link between mathematical skills and future socioeconomic status.

My research into the development of early mathematical skills showed that studies had been conducted mainly in early years settings, with many studies showing the positive benefits of play. This made me curious about the viability of play being used as an intentional pedagogy in early years contexts, and so I conducted a research project in Prep classrooms over a period of a few months.

The teachers who were part of the project had two interviews with me (one pre-lessons, one post-lessons). They also had a professional learning session with me to ensure that we all had a common understanding of play and the critical need to develop early maths skills, and they delivered three play-based lessons that I had written (with all resources included in a lesson box). I have included these lessons at the end of this chapter, if you are curious to learn more about them. They are intended to be an example of lessons that could be used in the classroom using a play-based pedagogy.

After delivering the three play-based lessons, the teachers reported that they, as educators, had fun delivering the lessons, and that they were surprised about the rigour embedded within the lessons. The teachers observed deep student engagement, and they felt that misconceptions were easily uncovered throughout the lessons, which enabled the teachers to target specific areas of understanding for specific students. Thanks to the planning framework that sat behind the lessons, teachers knew the content and goals of the lesson clearly, and they used play as an effective pedagogical tool to deliver and enhance learning.

One of the teachers articulated her shock at the depth of knowledge students demonstrated when they observed patterns between odd and even numbers as students went beyond the parameters of the lesson to extend and build on each other's thinking. Another teacher reported that her students naturally used a range of strategies to solve problems and demonstrate persistence, resilience, creativity, and critical thinking, without teacher prompting.

Here are some of the themes that emerged from the project, as reported by the teachers:

- Play is a strong foundation for learning.
- Play allows students to relax, have fun, be themselves, and learn.
- Play facilitates the growth of interaction and social skills.
- Play as a pedagogy facilitates discussions that enable students to feel comfortable sharing their thinking and learning.

- Teacher-directed play is a valuable pedagogy in the classroom because it effectively addresses the curriculum and meets learning outcomes.
- Using play as a pedagogy promotes critical thinking and natural extension and caters to the vast learning needs of the class (all students can contribute and experience success).

It is important to note that this research was conducted as a small-scale case study, so the findings of the study are not generalisable. However, the insights of the project, and the review of literature that was conducted, do shed light on the viability and rigour of play as a pedagogy in primary schooling.

Play as a pedagogical tool

Play as a pedagogy promotes the development of critical thinking, creativity, reasoning, metacognition, and social skills. Paatsch et al. (2024) argue that these are all essential skills in early mathematics learning and can be cultivated through collaborative, play-based experiences that invite discussion and reflection. Through play, students experience decreased maths anxiety and increased engagement. Learning in this way can enhance creativity, critical thinking, reasoning, discussion, and metacognition.

The lessons in the case study research project above were designed with consistent elements, which ensured the rigour of each lesson (summarised in Figure 16 on page 137):

- **Content Descriptions:** We need to work within the framework of the Australian (or state-based) Curriculum. When we plan lessons, we need to ensure that these lessons align with the content we are required to teach.
- **Early maths skills:** As part of the research project, I was focusing on the development of early maths skills. You can

read more about these skills in Chapter 3. I wanted to show the teachers in my project a clear alignment with the crucial maths skills that need to be developed by age seven.

- **General Capabilities:** Achievement Standards and Content Descriptions are not the only elements of the Australian Curriculum. Our curriculum is three-dimensional, and along with content and assessment, the lessons had a range of General Capabilities embedded.

- **Learning intentions:** It is essential for teachers to be clear on what they are teaching and why. The Australian Institute for Teaching and School Leadership (AITSL) states that learning intentions are 'descriptions of what learners should know, understand and be able to do by the end of a learning period or unit' (n.d., p. 1). Furthermore, meta-analysis research conducted by Hattie found that 'goals are important for enhancing performance' (Hattie, 2009, as cited in Department of Education and Training Victoria, 2017, p. 8).

- **Key guiding questions:** Along with the overarching learning intentions for the lesson, teachers need to be mindful of the questions that they ask. This is a skill that I am still developing after almost 20 years of teaching. It is something I strive to refine, because I am aware that my use of questioning to guide learning, extend learning, support learning, and undercover misconceptions in order to address gaps in understanding all rely on me asking the right questions at the right time. The better my questions, the better I can understand and address the learning needs of my students, and the better their outcomes will be. In Lyn Sharratt's research on teacher clarity, she emphasises the need to ask good questions to ensure effective teaching and learning (Clarity Learning Suite, 2024).

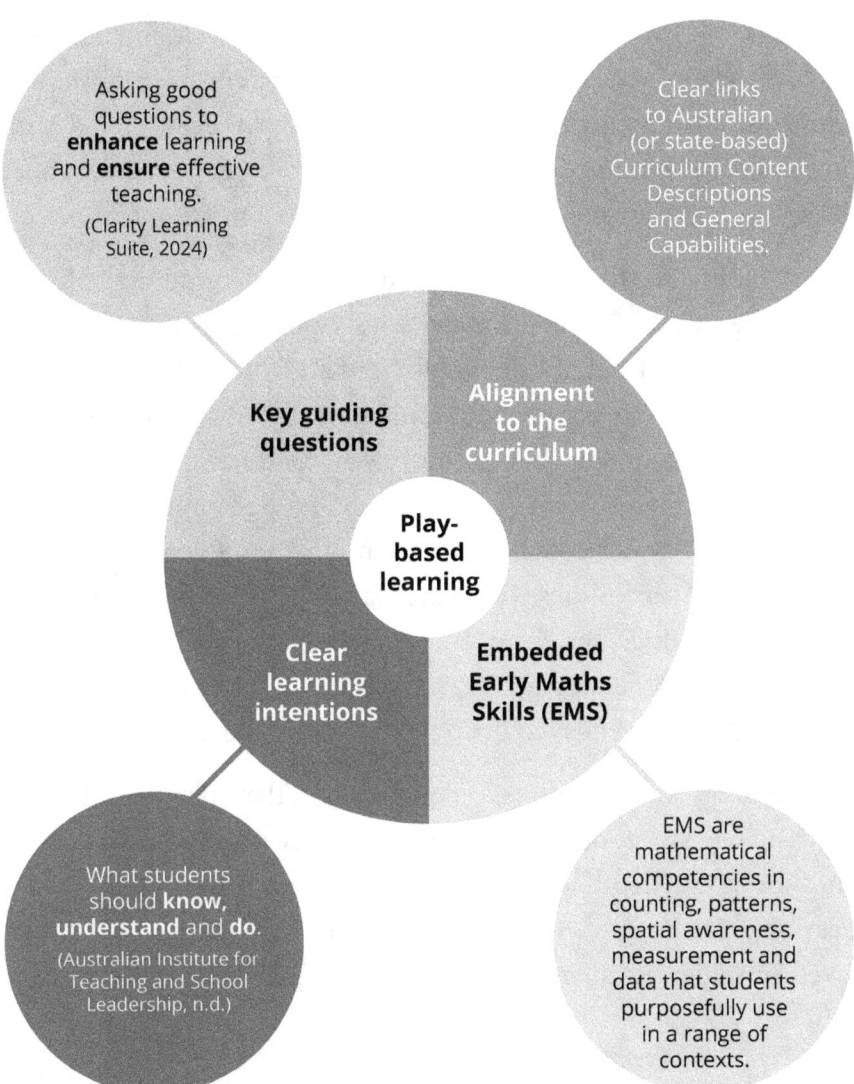

Figure 16. A summary of play-based learning as a pedagogy.

In an Australian project involving 19 kindergarten children, Cohrssen and Pearn (2021) investigated spatial thinking by setting students a play-based task of drawing the route from their home to school. The children were in their final few months of kindergarten and were actively engaged in school visits and transitions, ready to begin their Foundation Year of schooling the following year. Children were encouraged by their educators to engage in sustained conversations about their new schools. They were prompted with questions by educators to elicit their thinking about location, direction, and shape on the journey from home to school.

Every child in the project demonstrated a minimum of two Australian Curriculum Content Descriptions, with one child demonstrating 11 out of 13 possible Content Descriptions that relate to spatial thinking across the Foundation to Year 3 mathematics curriculum. Although this project was conducted with a small number of students in one kindergarten and does not give enough scope to draw wider conclusions, the project does demonstrate the wide range of capabilities with which students enter the Foundation Year of schooling. The project also demonstrates the possibilities that play-based pedagogies may have in mathematics, especially when students are engaged in contexts that are meaningful to them.

The findings of this project demonstrate that play-based tasks may enable all students to experience success in their learning and show the depth of their knowledge; and, as Cohrssen and Pearn (2021) suggest, they can 'meet children at their individual levels of understanding in order to authentically scaffold children's thinking' (p. 57).

The lesson plans

Lesson Plan 1: The Double Decker Bus	
Lesson Duration	45 mins
Content Descriptions	• quantify and compare collections to at least 20 using counting and explain or demonstrate reasoning – AC9MFN03 • represent practical situations involving addition, subtraction and quantification with physical and virtual materials and use counting or subitising strategies – AC9MFN05
Early Maths Skills	Counting and patterns
General Capabilities	Numeracy, Literacy, Social and Personal Capability, Critical and Creative Thinking
Learning Intention	We are learning to count to 20 using part-part-whole thinking.
Key Guiding Questions	• *Have you ever been on a bus before?* • *How many students are on the red deck?* • *How many students can you see on the blue deck?* • *How many students are on the bus altogether?* • *How do you know?*
Lesson Sequence	1. Ask the students if they have ever been on a bus before. Have students share their experiences. 2. Read the book *The Double Decker Bus* (Liu & Twomey Fosnot, 2008). 3. Have 10 chairs set up in 2 rows of 5 with a blue dot on each chair, a metre of space, and then another 10 chairs set up in the same way with a red dot on each chair. Have one chair at the front of the bus with a driver's hat and wheel ready. 4. Around the room there are 5 bus stops that students can choose to stand at – Cat Stop, Dog Stop, Turtle Stop, Bird Stop and Horse Stop. Give students the choice of which bus stop to stand at, while they wait for the bus.

Lesson Sequence (cont.)	5. Pretend to hop on the bus, put on the driver's hat, and pick up the wheel. Show the students how excited you are to be the bus driver.
	6. One stop at a time, call out, 'All aboard! Next stop, Cat Stop!' Have students enter the bus one at a time, and work as a class to count each person as they hop on the bus. Students can sit on the red deck or the blue deck (indicated by red dots or blue dots sticky-taped onto the chairs). Model counting the students on the red deck first, then the blue deck, and adding the two numbers together to work out how many people are on the bus. Once students help you with this, you can get everyone to put their 'seat belts' on and you can go to the next stop to collect some more students and repeat the modelling of counting and adding the two parts/numbers together to find out how many people are on the bus.
	7. Follow the lead of the students to either play again, have a student as a bus driver, or divide into two buses so that you have two drivers with 10 in each bus and play in smaller groups with the assistance of a teacher's aide.
Resources Needed	• Interactive panel, TV, or projector with laptop • *The Double Decker Bus* book • 20 chairs • 10 blue dots • 10 red dots • Sticky tape • Cat, Dog, Turtle, Bird and Horse Stop poles and signs • Driver's hat • Steering wheel

	Lesson Plan 2: Number Detectives
Lesson Duration	30 mins per small group
Content Descriptions	• quantify and compare collections to at least 20 using counting and explain or demonstrate reasoning – AC9MFN03 • partition and combine collections up to 10 using part-part-whole relationships and subitising to recognise and name the parts – AC9MFN04
Early Maths Skills	Counting and patterns
General Capabilities	Numeracy, Literacy, Social and Personal Capability, Critical and Creative Thinking
Learning Intention	We are learning that a total number can be made in many ways.
Key Guiding Questions	• *I wonder...* • *Do you have any ideas to help us solve the mystery of the Number Criminals?* • *How big is the Number Criminal?* • *Can you show me?* • *How do you know?* • *Can you make the number in another way?* • *How big will you make your Number Criminal?* • *Could you show the same number in another way?* • *Would the same number in another way make a different shape?* • *How do you know?* • *Show me your thinking using your Numicon pieces.*
Lesson Sequence	1. Set up the Number Criminals mat on a large flat surface like a table or the floor. Have the small group of students sit around the mat so that they can reach it easily from where they are sitting to be able to draw. 2. Say: 'Detectives are people who have a special job to look for clues and solve mysteries. Today, we have a special mystery to solve. Some Number Criminals have been captured, and I need your help to investigate how big each Number Criminal is.

Playful Maths 141

Lesson Sequence (cont.)	The only things left at the scene of the crime are these pieces.' Invite the students to tell you what they notice about the Numicon pieces (have one piece of each colour and number clearly shown so that students can see patterns and draw conclusions). Say: 'I wonder how these pieces might help us to solve the mystery. Does anyone have any ideas?' Invite students to share their thinking about how they might use these pieces to solve the mystery for each Number Criminal. 3. Next, invite students to work in pairs and use the Numicon pieces to find out how big each Number Criminal is. A key element to this activity is using questioning to support students in their thinking. Once students have found out how big they are, encourage them to describe how they found this out. Model language back to students like: 'Oh! You worked out that this Number Criminal had 4 altogether and you put 2 and 1 and 1 together to solve the mystery! Great work! I wonder if you could make 4 in another way.' 4. Once students have solved the Number Criminals in at least one way each, invite them to create their own Number Criminal using the Textas, Numicon pieces, and eye stickers. Have students solve each other's Number Criminal mysteries. 5. Over the next few days, leave some Numicon pieces, large sheets of paper, Textas, and eye stickers out for students to explore this concept further independently. Encourage their thinking and perhaps share their puzzles with others to celebrate their thinking. If you are able to, you may also like to make a wall of Number Criminals using the children's work.
Resources Needed	- Numicon set - Textas - Eye stickers - Butcher's paper - A3 paper

	Lesson Plan 3: Let's Build!
Lesson Duration	45 mins
Content Descriptions	• quantify and compare collections to at least 20 using counting and explain or demonstrate reasoning – AC9MFN03
Early Maths Skills	Counting and spatial awareness
General Capabilities	Numeracy, Literacy, Social and Personal Capability, Critical and Creative Thinking
Learning Intention	We are learning that 20 can be shown in many different ways.
Key Guiding Questions	• Can you lay a block down to make it flat so that you can put another block on top? • Could you do this another way? • Tell me about your design. • Explain how you made 20. • How could you make sure your tower is strong? • Is your tower a new design? • What makes your design special? • Can you explain how you designed your tower?
Lesson Sequence	1. Wear a hard hat and vest. Say: 'Prep, I have some exciting news. A special letter has been delivered to our classroom. Should we open it?' Open the letter and read it to the students. The letter invites each student to create a tower out of 20 blocks. Say: 'Engineers have a special job to do. Some engineers design houses. They make sure that the walls are strong, that the roof stays on, and that the floor is strong. Today we get to be engineers and we each get to design a tower. There are some special rules that we have to follow. First, we can only make one tower. Second, the tower can be as wide and as tall as you would like it to be. Third, you must first design your tower using 20 paper rectangles and show it to me to explain your design BEFORE you build your tower. Are you ready, engineers?!'

Lesson Sequence (cont.)	2. Use the design sheets and the paper rectangles provided and invite students to start placing their rectangles on their paper, without gluing them. Encourage them to make their design special and unique and encourage them to move their rectangles in different ways to help them picture different ways to make a tower of 20 blocks other than just straight up. Encourage them to stack, layer, and put blocks in either or both vertically and horizontally. Place a few blocks in the centre of each table for students to test out how the blocks sit and stack. 3. Once they have shown you their design and explained their thinking to you, they can stick the paper down and use this design to help them build their tower of 20 blocks. 4. Be sure to take a photo of each student with their design and tower and either show these designs on a digital slide show to celebrate the different ways that students used 20 blocks to make their tower or print them to make into a display wall at the students' height for them to admire the work of their peers.
Resources Needed	• Hard hat and vest • Large, printed letter resource with design specifications on it • 20 rectangular pieces of coloured paper for each student (provided) • 20 rectangular blocks of a consistent size per student (provided) • 1 student design sheet per student printed on light card

Lesson 3: Letter Resource

LET'S BUILD!

Dear Prep students,

A local tower builder needs your help to come up with a new design for a tower. Your tower can only have 20 blocks and it can be as wide or as tall as you would like it to be. Your tower needs to be able to stand up and be strong on a windy day, so be careful not to make it too tall!

Happy building,
Prep students!

From
The Tower Builders

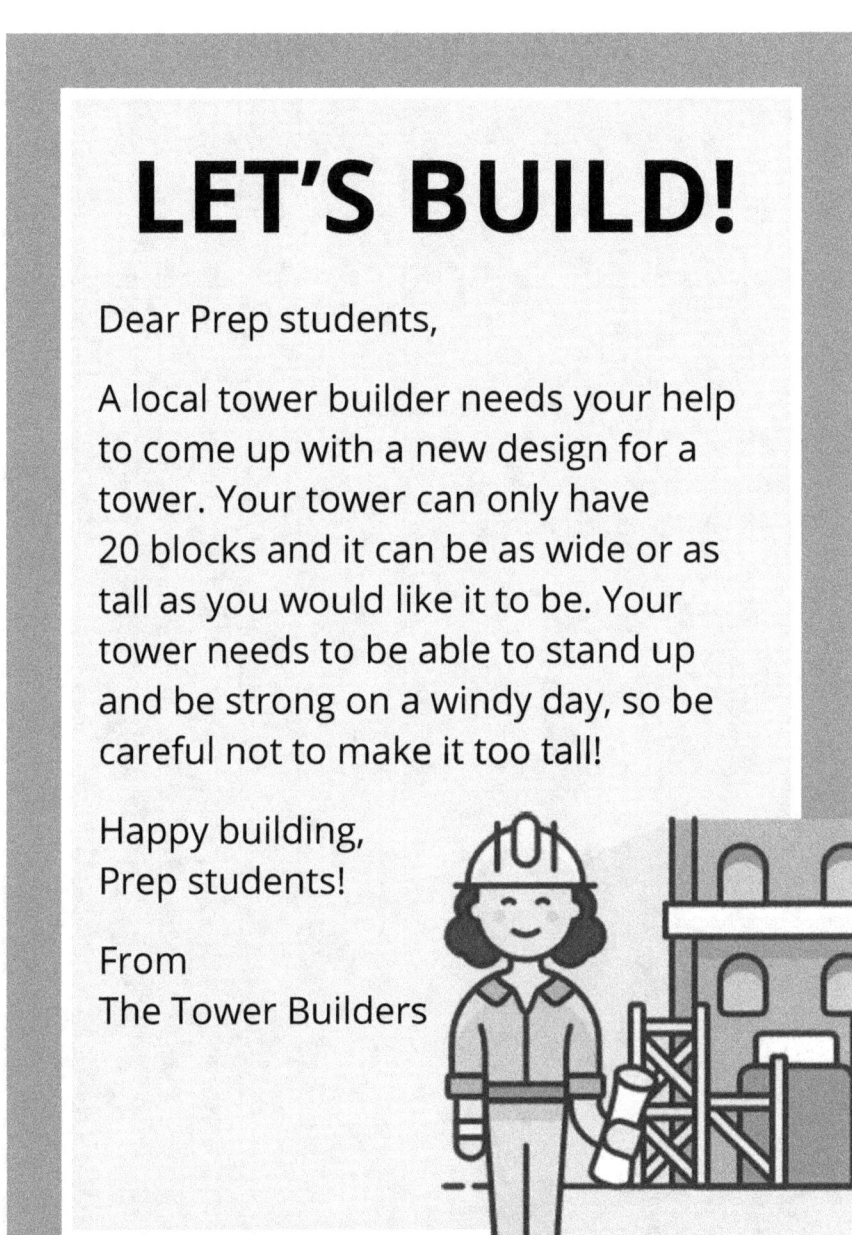

Lesson 3: Student Design Sheet

My 20 Block Tower

Name:

CHAPTER SUMMARY

- Through play, children learn resilience, persistence, problem-solving, collaboration, and more.

- Play is essential for cognitive, social, and emotional development.

- When students engage in playful learning, they are more likely to take risks, experiment with trial and error, make and learn from mistakes, and develop a sense of curiosity and joy in their learning.

- By embedding playful learning into the classroom, we can build classrooms where all students are engaged, challenged, and joyful in their mathematical learning journey.

- Play as a pedagogy promotes the development of critical thinking, creativity, reasoning, metacognition, and social skills.

9

Primary Mathematics Essentials

I consider myself privileged to be able to work with primary school-aged children. They bring such varied and rich experiences from their homes and families, and they enter school with such eagerness and curiosity. To see the joy and wonder on a small child's face when they discover something new is one of the most precious things to behold as a teacher. Throughout the primary years, teachers walk alongside students to help them to grow in their knowledge and skills, to encourage them as they develop resilience and persistence, and to facilitate engaging experiences that leave students hungry to know more.

The Sydney Children's Hospital Network (2025) describes the years of primary schooling as significant as children change physically, cognitively, socially, and emotionally. Throughout the primary schooling years, our students undergo rapid change and growth as they learn and increasingly develop independence. Piaget's theory of learning and development asserts that as children grow, they acquire more knowledge and grow in cognition as their capacity to reason, problem-solve, and make meaning increases (The Education Hub, 2021).

Throughout the primary years of schooling, children move through two of the four stages of development, according to Piaget's theory (The Education Hub, 2021). Children enter school during the *Pre-Operational Stage* of development, from approximately two to seven, when their thinking is dominated by visual images and symbolic thought, and responses are based on feelings, intuition, and sensorimotor experiences. Children in this stage of development have difficulty understanding abstract concepts and generalising across contexts.

Students spend most of their primary schooling years in the next stage of development, the *Concrete Operational Stage,* from approximately seven to 12. In this stage of development, children think more logically and begin to generalise across contexts and understand concepts more deeply. Throughout this stage of cognitive development, children move beyond egocentricity and perception-based thinking towards thought processes that are more flexible and structured.

As teachers, we know how much the students in our primary school grow and change. We see their little faces changing over time, their height increasing, and their self-control growing. But how often do we stop to reflect on not only the extensive growth that takes place in the primary schooling years but the significant shifts in cognitive development that occur? We need to be responsive to not only our students' academic learning needs but to the stage of cognitive development that they are moving through. In primary schools, we need to support students in the early years of schooling with plenty of visual and symbolic examples that are connected to their lives. We then need to support them as they begin to generalise across contexts and eventually move into the next stage of deeper understanding as they begin to think more abstractly.

Concrete materials

When we consider the information throughout this book, including the information about child development above, it is clear that our students need to be given a range of learning experiences to help them to grasp and understand concepts. Concrete materials, also known as manipulatives, are important for mathematical teaching and learning.

Clements and Sarama (2018) argue that concrete materials are essential for mathematics teaching in some contexts and helpful in others. For example, in the early years of primary schooling, children need physical objects to count to be able to understand quantities. Young children need to use concrete materials to physically act out addition and subtraction problems so that they can literally *see* changes to a quantity when items are physically being counted, added, or taken away. Remember, these little people are in a stage of cognitive development where they are not yet able to generalise but rely on visual representations to help them to understand.

Jerome Brunner developed a theory of cognitive development, called the CRA Model (Mathematics Hub, 2024). This model is a useful framework to help teachers understand how students progress in their understanding of mathematical concepts. The CRA Model includes three key stages of mathematics learning:

1. Concrete
2. Representational (pictorial)
3. Abstract.

Students begin by using concrete materials (such as counters, Unifix cubes, blocks, or measuring tools) which allow them to physically manipulate and experience mathematical concepts. The concrete stage is crucial, as it builds the foundational understanding, making the abstract concepts more visible and tangible. Next, students move into the pictorial stage, where they represent mathematical concepts using diagrams, drawings, or models. Then finally, in the third stage of learning, students use numbers and symbols to represent the mathematical concepts in an abstract way. Each of these stages of learning is important in supporting students to build both understanding of and fluency in mathematics.

Sometimes, in our rush to cover the breadth of content in the curriculum, we skip past stages and get frustrated, wondering why our students aren't listening properly to what we are teaching. We may skip the concrete stage or perhaps use concrete materials and not address the abstract stage. We wonder why our students are staring blankly at us, asking lots of questions repeatedly, or are overly reliant on us and not showing independence. Often it is because we haven't allowed them to understand the mathematical concept at each stage and, as a result, they don't fully grasp the concept.

I have witnessed some early years classrooms with an overreliance on concrete materials and a hesitancy to move students into the abstract stage of their mathematics learning, and I have witnessed plenty of upper primary and high school classrooms where students are expected to go straight to abstract thinking without foundational understanding from concrete and pictorial experience. In both scenarios, many students are left confused and do not deeply understand the concepts being taught.

At my school, my teachers have a literal 'toolbox' to rely on in every classroom, from Prep to Year 6. We call this a 'maths trolley':

a simple two-tier metal cart on wheels stocked with a range of concrete materials. Its portability allows teachers to easily access and use hands-on resources during lessons to model concepts, present ideas in different ways, or provide targeted support to students who may be struggling. By engaging students with manipulatives in the concrete phase of mathematical learning, the trolley helps build strong foundational understanding.

An example of the use of a range of concrete materials is when students are learning about numbers up to 100. A deep knowledge and understanding of numbers up to 100 is essential not only in Year 1, in accordance with the Australian Curriculum, but as a foundational concept that is built on in later primary as decimal places and financial mathematics involving cents and percentages are taught. Without a solid understanding of numbers up to 100, learning these later mathematical concepts can be exceptionally difficult.

Here are some of the concrete materials I would use to teach numbers to 100:

- Movable 100s board
- Giant 100s board on the floor with movable number pieces
- Printable 100s boards and transparent counters
- Johnstone Number Line
- Linking chains
- Paddle Pop sticks and rubber bands
- MAB blocks
- Counters
- Unifix cubes
- Magnetic dots
- Bead strings
- Number lines
- 1 metre ruler
- Place value charts.

By having some of these items on hand in your maths trolley, you are able to pivot and respond to the learners in front of you. You may be using MAB blocks and noticing that your learners are finding it difficult to understand how ten tens are the same as 100. You could use another tool, such as a Johnstone Number Line, to physically show how one ten is made up of ten ones, and ten tens are made up of 100 ones or make the number 100 in total. You may then use bead strings to reinforce this concept as students physically make numbers up to 100 by physically moving the beads.

Concrete materials are essential for the foundational mastery and understanding of mathematical concepts. They help students to form the first and most critical layer in understanding a concept, and from this understanding students are able to more easily grasp concepts in symbolic and abstract ways. Concrete materials are useful across the primary years of learning and into the high school years and are invaluable tools that support conceptual understanding in mathematics.

Sample maths trolley items for lower primary

- 1 x container of 20 plain coloured, round magnets
- 3 x containers of assorted counters
- 100 x Unifix cubes
- 2 x sets of playing cards
- 1 x large foam dice
- 2 x large pocket dice
- 1 x set of linking chain pieces
- 3 x writable and erasable number lines to 20
- 1 x large Rekenrek
- 5 x small Rekenreks
- 2 x magnetic, writable numeral expanders
- Pencil case of 100-bead strings
- Pencil case of magnetic tens frames
- Rubber bands
- Paddle Pop sticks.

Sample maths trolley items for middle primary

- A small selection of MAB blocks
- 3 x containers of assorted counters
- Various sets of maths card games
- Large number line to 100
- Pencil case of coins
- Containers of assorted dice
- 100 x Unifix cubes
- Pencil case of 100-bead strings
- 2 x magnetic, writable numeral expanders
- 1 x container of operations dice
- 1 x class set of measuring tapes
- Pencil case of tangram shapes
- 1 x large pocket dice.

> **Sample maths trolley items for upper primary**
> - Johnstone Number Line
> - Flexible 1 metre ruler
> - Calculators
> - A small selection of MAB blocks
> - 1 x container of counters
> - Various sets of maths card games
> - Large number line to 100
> - Pencil case of coins
> - Containers of assorted dice (standard, operations, 10-sided, 20-sided, etc.)
> - 100 x Unifix cubes
> - Pencil case of 100-bead strings
> - 2 x magnetic, writable numeral expanders
> - 1 x container of operations dice
> - 1 x class set of measuring tapes
> - Pencil case of tangram shapes
> - 1 x large pocket dice.

Numeracy picture books

There are a range of excellent books that bring mathematical concepts to life for our students. In lower primary, many books encourage counting and positional language. One of my all-time favourite books for supporting students to understand and use positional language is *Rosie's Walk* by Pat Hutchins. Along with reading this book, I like to set up an invitation to play on a table with a simple farm set up. Students can play with this and role-play moving Rosie and the fox over, under, through, and around items from the story. Using three-dimensional objects and animals really brings positional language to life for students.

A book that I love to read to students in middle primary to reinforce place value is *Sir Cumference and All the King's Tens* by Cindy

Neuschwander. This book is an excellent example of using visual representations to show the magnitude of size as numbers get larger. It is a good book to launch maths investigations about place value and to explore the concept of the size of numbers (e.g., *How big is 1000?*).

Finally, my favourite book showing a range of statistics using infographics and interesting illustrations is *If the World Were 100 People*, written by Jackie McCann and illustrated by Aaron Cushley. This thought-provoking, beautifully illustrated book is an excellent introduction to statistics for primary school-aged children. It explores a range of statistics about people around the world and will provoke deep discussions and further investigations in your classroom.

I share books that I have used and recommend on my Instagram page, so head to @Primary_Pedagogy to see some of the latest recommendations that may be suitable for your class or year level.

The importance of hands-on learning experiences

'Tell me and I forget; teach me and I remember; involve me and I learn.'

This quotation, sometimes attributed to Benjamin Franklin but also said to stem from a Chinese saying, is often used in professional learning conferences to inspire teachers to take action. I have to say that I agree with the sentiment of this quotation, because learning is not static – it is dynamic – and to truly understand something we need to be able to engage with it in a range of ways. This is true for all subject areas. We need time to listen, process, discuss, ask questions, and engage in a range of activities to be able to grasp new concepts. This multi-dimensional engagement with learning occurs across the curriculum and across all years of schooling. Involving students in the learning journey is what makes teaching the amazing career that it is.

Engagement in mathematics learning is strengthened when students have the opportunity to physically engage with the content. It may be a teacher making use of marking cones to visually demonstrate the properties of triangular numbers, providing concrete materials to support students to solve a problem, or giving students paper to fold to uncover the fractional parts of a shape. It is important to provide a range of hands-on learning experiences, as this allows students to become active participants in their learning.

Beyond the primary schooling years, mathematics becomes increasingly complex and abstract. Concepts across the curriculum are both spiralled (where concepts are introduced and revisited throughout the curriculum) and sequential (building on foundational knowledge, progressively moving to more complex concepts). It is important that we provide time to allow students to investigate and explore concepts to support them to develop deeper understanding.

Learning is further enriched when the experience connects to the real lives and contexts of our students. Connecting learning to the contexts that students can relate to supports them in generalising their learning across contexts and seeing the real need for mathematics in everyday life. By providing hands-on learning experiences, we give students opportunities to try out their theories, to make mistakes in a context where they can make changes to their solution and try another method to solve the problem, and we create opportunities for collaborative learning.

Journalling

Maths journals are very important in our school. Every student has one. Every teacher has one. Our maths journals are special spaces because no one tells you what to write in these journals. Students can choose to draw, write, graph, draw diagrams, use symbols, or use numbers. The purpose of these books is to have a space where students 'think about their thinking'. It is a space for them to reflect on what they have learnt in a lesson, series of lessons, or unit of work to show what they understand. Teachers always prompt students with a question to structure an activity and to ensure that there is clear intentionality to the activity, but students are able to use this journal to display their thinking and understanding in whatever way suits them best.

Hattie and Zierer (2018) conducted over 1400 meta-analyses of evidence from 80,000 research studies, involving over 250 million students, to measure the effect of over 250 influences on student learning and achievement. Of these influences, 'metacognitive strategies' (Hattie & Zierer, 2018, p. 78) are among the most impactful on student learning. When metacognitive strategies are employed, students are transformed from passive consumers to learners who question, discuss their thinking, reflect on their mistakes, and are more willing to take on feedback. Learning becomes more dynamic, and the teacher is able to be responsive to student and planning needs as student learning becomes visible.

Maths journals not only provide a valuable space for students to think about what they have learnt in a lesson; they are also a powerful way for teachers to receive feedback on the effectiveness of what they have taught. By walking around the classroom while students journal, teachers can quickly gauge the impact of their lesson. They can see student misconceptions that need to be addressed; they can see who needs further support in learning a concept and who could be encouraged to go deeper with their learning. As stated earlier in this book, being a reflective practitioner is crucial to being a great teacher. We need not only to ensure that our students learn, but also to take the time to reflect on how well we have taught something, and to be willing to refine and adjust our teaching for the benefit of our students.

The photograph in Figure 17 shows the work of a Year 4 student who had been asked to share her maths journal example with the class and talk through her thinking. All students were given journalling time to reflect on their understanding about odd and even numbers. This student wrote the rules she knew, an example sum, diagrams, a short written explanation, and a final diagram proving how she knew that 8 was an even number. (Note: the photo was taken mid-way through the demonstration, so the student had not quite finished the last diagram she was drawing.)

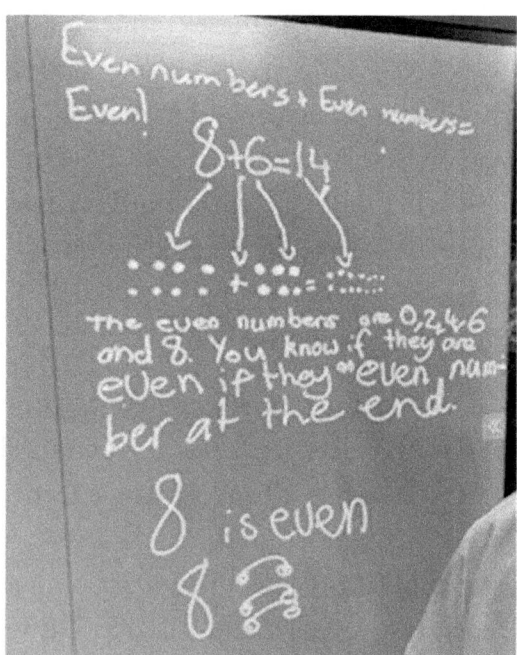

Figure 17. A photo of a student's maths journalling work explaining her understanding of even numbers.

The maths journals in our lower primary are A5 scrapbooks, while the journals in our upper primary are grid books (students may glue a piece of blank paper into their grid book if they choose to). This is part of book pack requirements, and journals are accessible for students from day one of the school year. I encourage teachers to showcase student thinking, creativity, and problem-solving. Teachers do not model 'how to' complete maths journals; rather, they celebrate the thinking of their students and ask students permission to share their insights, diagrams, or perhaps their page of work with the class. The students enjoy the freedom to communicate their thinking and understanding in a book that has no set rules, and they like to celebrate the thinking and insights of their peers.

Connections between home and school

As a parent, I value teachers who take the time to reach out and make meaningful connections to our family and home. Over the years, these experiences have included my children bringing a special bear home with a suitcase of clothing and a diary to write in about what the bear experienced in our home. In upper primary, my children were tasked with creating a family tree, and over a couple of weeks we visited family members' homes and encouraged our children to make phone calls, and our children each created a family tree – from this experience we also received wonderful family history stories, photocopies of old photographs, and artifacts (like old wooden board games) that grandparents and great-grandparents used in the past. As my children are now in high school, I have appreciated the teachers who have emailed at the beginning of the term to explain the core content for the term and offer any support. Home school connections are invaluable to our family.

Parents are central to schooling. They are the people who invest in their children in the years before school, do the evening routines and morning struggles, and know their children best. As discussed in Chapter 5, parents may have experienced maths anxiety and/or negative maths experiences in the past, and if this is the case they are likely to unconsciously perpetuate negative myths about mathematics. We need to provide families with positive mathematics experiences and, in my opinion, one of the most effective ways to do this is through maths backpacks (see Figure 18).

Maths Backpack Checklist

- A backpack
- Clear instructions about the activity or game
- Any accompanying resources
- A blank scrapbook for families to journal in.

Optional extras:

- A polaroid camera for families to take photos of them involved in the activity
- Stickers
- Sticky tape
- Glue
- Stamps
- Ruler.

Figure 18. A suggested list of items to include in a Maths Backpack.

Maths backpacks are simply backpacks containing maths activities and games that are sent home regularly for families to enjoy. They do not need to be expensive or fancy (you could even ask a family to donate an old backpack for you to use). All you need is a backpack, clear instructions about the activity or game, any accompanying resources, and a blank scrapbook for families to journal in. As a classroom teacher, I sent home the maths backpack twice a week, so that families had a few days to enjoy the backpack. On the scheduled return day, the student would bring the backpack back to class, and I set aside time in the daily class schedule for them to share their journal page. This page often contained photos of the family playing the game, along with other interesting things like strategies each player tried, tally marks for who won each round, notes and diagrams for problem-solving, hand-drawn pictures to explain how the family approached the task, and more. These were such rich moments of sharing and reflection in my classroom, and students couldn't wait to take the backpack home to their own family.

We also set up norms for use and had clear rules around how we would care for the items in the backpack. I wanted everyone to enjoy the items, but I also value teaching students how to care for the property of others. As this backpack belonged to our whole class, students took on the role of caring for the backpack and its items very seriously. They would sharpen any pencils, keep small items away from young children and pets, clean the kitchen table before unpacking the items, and not consume food while using the items. I am also realistic that accidents happen, and in the rare event that something broke or went missing, students were honest and apologetic with me and often created a solution themselves with their family so that the backpack could continue to be passed on to the next family, but perhaps with an adjustment to one of the resources.

Some of the items I like to include in maths backpacks are:

- Numeracy picture books with prompting questions and a linked investigation
- Maths games from around the world (I particularly sought out games that reflected the cultural backgrounds of students in my classroom)

- Simple board games
- Mancala – this is one of my favourite games to develop strategic thinking skills
- A printed-out game, with a few counters, inside a plastic sleeve (or laminated)
- Objects to sort
- Investigation prompts with simple materials
- Maths journal – a large scrapbook or display folder with plenty of different types of paper, pencils, stamps, stickers, and a ruler; all stored in a large pencil case inside the bag.

Recently, my school was awarded statewide Gold status for our active promotion, professional development, and use of mathematics in our school. We are one of only two primary schools in the state of Queensland in the past year to receive this recognition. To celebrate this milestone, I ran a competition over the Easter holiday break. I encouraged families to take a photo of themselves using maths in their everyday lives. Students had the chance to win a family board game as a reward. We were blown away by the many entries as parents sent in photos of students counting sausages when cooking on a campfire, calculating money when shopping, setting up stock in market stalls, building miniature houses with furniture out of recycled materials, graphing the number of M&M's in a bag, sorting and categorising collections of Cubeez, and more. We celebrated these contributions at our school assembly, and classroom teachers actively promoted the competition and highlighted ways that students used maths in the classroom, in the playground, and around the school.

There are many ways to create opportunities for families to engage in mathematics with their child. We need to be mindful of our students and context and create positive experiences for parents and their children. If we provide the ideas and resources, families generally embrace the activities that we provide and get on board.

Our students certainly are the greatest beneficiaries as they eagerly embrace positive connections between home and school.

Having fun with maths

Mathematics learning is essential for preparing students for their future, but that doesn't mean it has to be dull or uninspiring. While repetition plays a critical role in helping students understand mathematical concepts (particularly when it comes to building vocabulary), repetition doesn't have to be monotonous. By injecting fun into our teaching and providing students with a variety of ways to interact with mathematical ideas, we can make learning more meaningful and memorable.

One powerful way to make mathematics learning enjoyable and effective is through the use of maths games. Games allow students to practise key skills in a low-pressure environment, encouraging strategic thinking and reinforcing fluency and proficiency. They also create opportunities for collaboration and friendly competition, making maths a more social and dynamic subject. Alongside games, school-wide maths competitions can help foster a culture of excitement and positivity around mathematics, encouraging students to challenge themselves and celebrate their successes.

When choosing games, it is important that you, as the teacher, have clear intentions for playing the game. Giving students a game that does not reinforce a concept, or is not well written to help reinforce and practise the learning of a concept that has been explicitly taught in class, can create and reinforce misconceptions. Before you select or design a game for use in your classroom, ask yourself:

- What is the underlying learning intention of this game?
- What concept knowledge or skills do I expect to be reinforced when my students play this game?

When selecting games to be used to support learning, you need to check the rigour of the game, just like any other resource that you introduce into your classroom. The best place to start when selecting credible sources is the Australian Curriculum. Here are some steps to follow to help you determine the alignment of a maths game to your year level and students (summarised in Figure 19):

1. Does the game align with a specific Content Description in the Australian Curriculum for your year level? To check this, highlight key words and check any words you need further clarification of in the glossary.

2. If the answer is yes, does the content align with the key ideas shown as examples in the Content Description? An addition game can take many forms – check to see that the numbers used align with your year level (e.g., do you need games with one- and two-digit numbers, or three- and four-digit numbers?).

3. If the answer is yes again, ask yourself: What is the learning intention of this game for my students? Is the aim to practise and build fluency in a particular concept, or is it perhaps to learn ways to strategise when solving problems? Be clear about the intention of the learning behind the game.

4. If you have answered yes to all of the above, I recommend that you create a label for the game (if it is your own) or that you start a spreadsheet where you record which games align well with the curriculum year level you are teaching along with the learning intention of the game. This way, you will build a bank of resources to use throughout the years. This will help when you need to choose games to differentiate as you can make notes and categorise games for students who may need support, or perhaps it may be a suitable game for students who need to be extended.

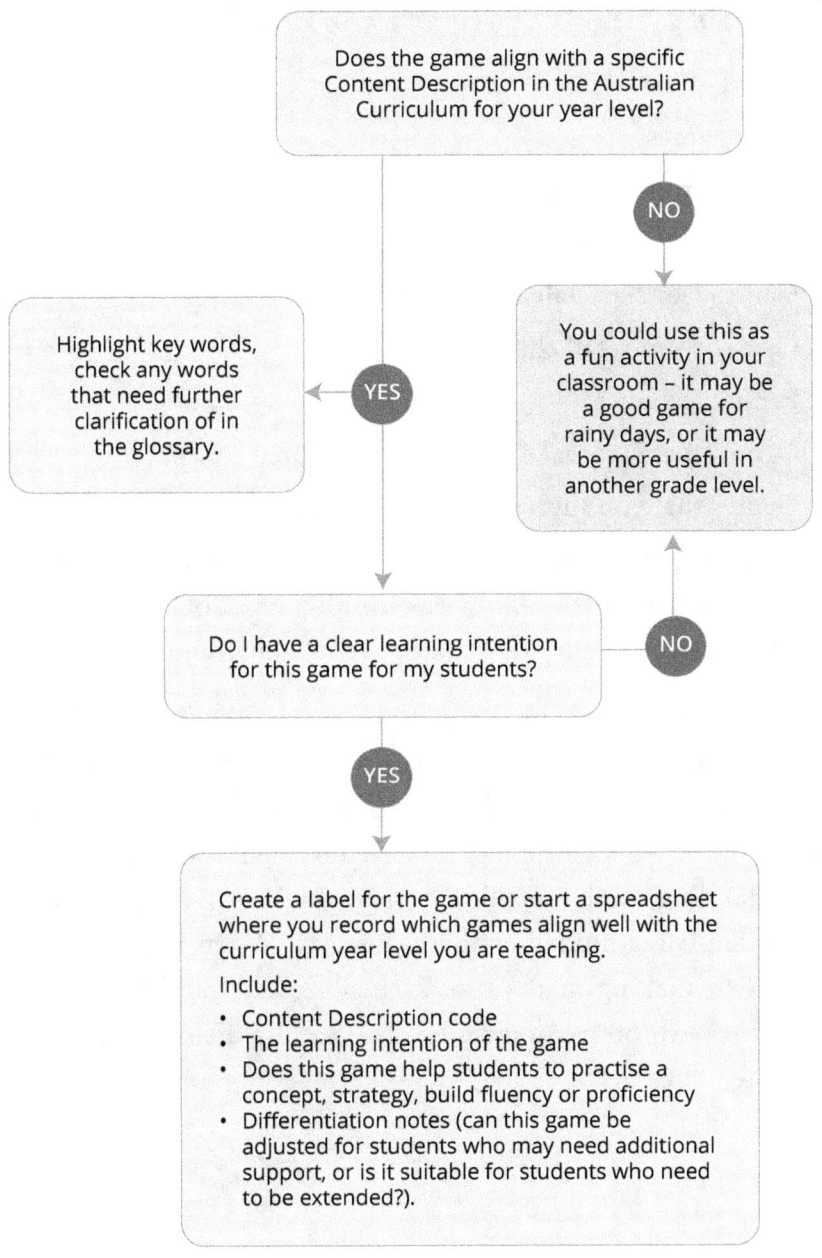

Figure 19. A flowchart to assist teachers in selecting quality, aligned resources for their classroom.

The types of games that you could use in your classroom include:

- Board games
- Card games
- Games from around the world
- Games from educational suppliers
- Games from specialised game stores
- Games from retail chain stores
- Printable games
- Games that you make yourself to reinforce concepts
- Games that you support students to create
- Computer games (there are so many choices, but again, go through the list above to check the alignment of the game or app before you subscribe to or purchase anything).

Promoting a positive maths culture in schools also involves the attitudes and actions of educators. Teachers play a crucial role in shaping how students perceive mathematics. By embracing a growth mindset, continuing to develop their own mathematical understanding, and modelling a love for learning, teachers can inspire similar attitudes in their students. When we, as teachers, commit to making maths fun, accessible, and valued, we set our students up for both immediate success and a lifelong appreciation of mathematics.

CHAPTER SUMMARY

- Teachers need to be responsive not only to students' academic learning needs but to the stage of cognitive development that they are moving through (Pre-Operational Stage – two to seven years; Concrete Operational Stage – seven to 12 years).

- The CRA Model (Concrete, Representational, Abstract) shows how students progress in their understanding of mathematical concepts.

- A 'maths trolley' allows teachers to easily access and use hands-on resources during lessons to model concepts, present ideas in different ways, or provide targeted support to students who may be struggling.

- Concrete materials are essential for the foundational mastery and understanding of mathematical concepts.

- Numeracy picture books help to bring mathematical concepts to life for our students.

- Engagement in mathematics learning is strengthened when students have the opportunity to physically interact through hands-on learning.

- Journalling encourages students to engage meta-cognitively. Maths journals are a valuable space for students to think about what they have learnt in a lesson; they are also a powerful way for teachers to receive feedback on the effectiveness of what they have taught.

- Teachers should create meaningful home–school connections to create positive experiences for parents and their children. Our students are the greatest beneficiaries, as they eagerly embrace positive connections between home and school.

- In order to select games, activities, books, and resources for their classroom, teachers should check that they explicitly align with the curriculum for the particular year level. Teachers also need to create clear learning intentions for the resource, game, book, or activity before using it in their classroom.

Conclusion

Mathematics teaching is complex and dynamic. Our own experiences as learners may impact how we teach mathematics. It is essential for teachers to reflect on their own experiences so that they can consciously and intentionally teach maths in positive and impactful ways.

The current landscape in our country is one where AEDC data shows an increase in students entering primary school with developmental vulnerabilities in the domain of language and cognitive skills. Nationally, there has been a decline in PISA maths results for our 15-year-old students since 2002, particularly for our lowest-achieving students. Compounding this decline, there is a decrease in enrolments in university STEM courses, which may put Australia in a difficult position in the future.

Research shows that students' maths skills at age seven may predict their socioeconomic status in middle age. Ensuring our students have a strong understanding of mathematics is critical both now and into their futures. This is reinforced by the integration of Numeracy in the Australian Curriculum as a General Capability that is built upon throughout each year of schooling to ensure that our students can use mathematics in their daily lives as active citizens in our society.

In our classrooms, we need to engage our students using the three dimensions of cognitive, behavioural, and emotional engagement.

These aspects of engagement are particularly important to support girls to feel more connected to mathematics. Furthermore, we need to create classroom cultures in mathematics lessons that encourage students to discuss their thinking, ask questions, and dig deeper to explore mathematical concepts.

Maths anxiety is a very real thing for teachers, parents, and students. It can manifest through psychological and cognitive symptoms which may include an increased heart and breathing rate, negative thoughts, worrying, or rumination. Teachers need to have an awareness of maths anxiety and support students to feel safe and supported in the mathematics classroom. School leaders also need to be aware of maths anxiety and support teachers in their own learning around mathematics.

There are various of ways that teachers can make a difference in their classroom and address the realities outlined above. Some ways that teachers can support their students include:

- Valuing and promoting creativity and curiosity
- Asking considered questions
- Using rich tasks and maths investigations
- Making connections to the real life of students
- Using the CRA (Concrete, Representational, Abstract) Model to support students to develop a deep understanding of mathematical concepts
- Using a range of concrete materials
- Integrating numeracy picture books into learning
- Using maths journals for students to reflect on their thinking and understanding
- Facilitating positive connections between home and school.

I hope this book has deepened your understanding of mathematics education in primary school settings. As you reflect on the ideas and strategies shared, I encourage you to set a practical, actionable goal to enhance your teaching. Remember, as a primary school teacher, you are a powerful agent of change – shaping young minds and inspiring a love of learning every day. Thank you for your dedication and for the incredible impact you have on your students. May this book support your continued growth and help you refine your practice to make an even more meaningful difference in your students' mathematical journeys.

References

Anders, Y., & Rossbach, H. (2015). Preschool teachers' sensitivity to mathematics in children's play: The influence of math-related school experiences, emotional attitudes, and pedagogical beliefs. *Journal of Research in Childhood Education, 29*(3), 305–22. https://doi.org/10.1080/02568543.2015.1040564

Attard, C. (2022). *Mathematics for contemporary learners – Lecture 1.* Western Sydney University.

Australian Association of Mathematics Teachers. (2009). *School mathematics for the 21st Century: Some key influences.* AAMT Inc.

Australian Curriculum. (2022). *Critical and Creative Thinking.* https://v9.australiancurriculum.edu.au/f-10-curriculum/general-capabilities/critical-and-creative-thinking?element=0&sub-element=0

Australian Curriculum. (2024). *F-10 Curriculum overview.* https://v9.australiancurriculum.edu.au/f-10-curriculum/f-10-curriculum-overview#3

Australian Curriculum. (2025). *Understand this learning area: Mathematics.* https://v9.australiancurriculum.edu.au/curriculum-information/understand-this-learning-area/mathematics

Australian Curriculum, Assessment and Reporting Authority [ACARA]. (2024). *School students with disability.* https://www.acara.edu.au/reporting/national-report-on-schooling-in-australia/school-students-with-disability

Australian Government, Department of Education. (2019). *The Alice Springs (Mparntwe) Education Declaration.* [PDF]

Australian Government, Department of Education. (2022). *Belonging, being and becoming: The early years learning framework for Australia (V2.0).* https://www.acecqa.gov.au/sites/default/files/2023-01/EYLF-2022-V2.0.pdf

Australian Government, Department of Education. (2024). *Concrete, representational, abstract (CRA).* Mathematics Hub. https://www.mathematicshub.edu.au/plan-teach-and-assess/teaching/teaching-strategies/concrete-representational-abstract-cra/

Australian Institute for Teaching and School Leadership. (n.d.). *Learning intentions and success criteria strategy.* https://www.aitsl.edu.au/docs/default-source/feedback/aitsl-learning-intentions-and-success-criteria-strategy.pdf

Bicer, A., Chamberlin, S., & Perihan, C. (2020). A meta-analysis of the relationship between mathematics achievement and creativity. *Journal of Creative Behaviour, 55*(3), 569–90. https://doi.org/10.1002/jocb.474

Bicer, A., Marquez, A., Colindres, K., Schanke, A., Castellon, L., Audette, L., Perihan, C., & Lee, Y. (2021). Investigating creativity-directed tasks in middle school mathematics curricula. *Thinking Skills and Creativity* (40). https://doi.org/10.1016/j.tsc.2021.100823

Boaler, J. (2016). *Mathematical mindsets: Unleashing students' potential through creative math, inspiring messages and innovative teaching*. Jossey-Bass.

Boaler, J. (2022). *The elephant in the classroom: Helping children learn and love maths* (Rev. ed.). Profile.

Boaler, J., Munson, J., & Williams, C. (2020). *Mindset mathematics: Visualising and investigating big ideas*. Jossey-Bass.

Cambridge University Press. (2025). *Cambridge dictionary: Meaning of numeracy in English*. https://dictionary.cambridge.org/dictionary/english/numeracy

Carey, E., Devine, A., Hill, F., Dowker, A., McLellan, R., & Szűcs, D. (2019). *Understanding mathematics anxiety: Investigating the experiences of UK primary and secondary school students*. Centre for Neuroscience in Education, University of Cambridge. https://doi.org/10.17863/CAM.37744

Chamberlin, S., & Mann, E. (2021). *The relationship of affect and creativity in mathematics: How the five legs of creativity influence math talent*. Routledge.

Clarity Learning Suite. (2024). *Alignment of the CLARITY work with the explicit teaching model*. https://cls.claritylearningsuite.com/alignment-of-the-clarity-work-with-the-explicit-teaching-model/

Clements, D. H., Lizcano, R., & Sarama, J. (2023). Research and pedagogies for early math. *Education Sciences, 13*(8), 839. https://doi.org/10.3390/educsci13080839

Clements, D. H., & Sarama, J. (2018). Myths of early math. *Education Sciences, 8*(2), 71. https://doi.org/10.3390/educsci8020071

Clements, D. H., Sarama, J., & Germeroth, C. (2016). Learning executive function and early mathematics: Directions of causal relations. *Early Childhood Research Quarterly, 36*, 79–90. https://doi.org/10.1016/j.ecresq.2015.12.009

Cohrssen, C., Church, A., Ishimine, K., & Tayler, C. (2013). Playing with maths: Facilitating the learning in play-based learning. *Australasian Journal of Early Childhood, 38*(1), 95–99. https://doi.org/10.1177/183693911303800115

Cohrssen, C., & Pearn, C. (2021). Assessing preschool children's maps against the first four levels of the primary curriculum: Lessons to learn. *Mathematics Education Research Journal, 33*(1), 43–60. https://doi.org/10.1007/s13394-019-00298-7

Commonwealth of Australia. (2025). *The Australian Early Development Census: An Australian Government initiative*. https://www.aedc.gov.au/

Condon, J. (2002). *If the world were 100 people*. (Illustrated by S. Dillingham.) Kids Can Press.

Deiss, H. (2020, Feb 24). *NASA: Katherine Johnson: A Lifetime of STEM*. https://www.nasa.gov/learning-resources/katherine-johnson-a-lifetime-of-stem/

Department of Education and Training Victoria. (2017). *High impact teaching strategies: Excellence in teaching and learning*. https://www.education.vic.gov.au/Documents/school/teachers/management/highimpactteachingstrat.pdf

Department of Education and Training Victoria. (2020). *Mathematics anxiety*. https://www.education.vic.gov.au/Documents/school/teachers/teachingresources/discipline/maths/MTT_Mathematics_Anxiety.pdf

DePascale, M., Butler, L. P., & Ramani, G. B. (2023). The relation between math anxiety and play behaviors in 4- to 6-year-old children. *Journal of Numerical Cognition, 9*(1), 89–106. https://doi.org/10.5964/jnc.9721

Doctoroff, G. L., Fisher, P. H., Burrows, B. M., & Edman, M. T. (2016). Preschool children's interest, social-emotional skills, and emergent mathematics skills. *Psychology in the Schools, 53*(4), 390–403. https://doi.org/10.1002/pits.21912

Dowker, A., Sarkar, A., & Looi, C. Y. (2016). Mathematics anxiety: What have we learned in 60 years? *Frontiers in Psychology, 7*, Article 508. https://doi.org/10.3389/fpsyg.2016.00508

Evidence for Learning. (2023). *Guidance report: Improving mathematics in the early years with children aged 3–7 years*. https://evidenceforlearning.org.au/education-evidence/guidance-reports/early-maths

Foundation for Young Australians. (2017). *The new work smarts: Thriving in the new work order*. https://www.fya.org.au/app/uploads/2021/09/FYA_TheNewWorkSmarts_July2017.pdf

Grace, G. (2023). *Queensland turns the page on reading in schools*. https://statements.qld.gov.au/statements/98983

Hattie, J., & Zierer, K. (2018). *10 Mindframes for visible learning: Teaching for success*. Routledge.

Henry, M. (2015). Learning in the digital age: Developing critical, creative and collaborative skills. In S. Younie, M. Leask, & K. Burden (Eds.), *Teaching and learning with ICT in the primary school* (2nd ed.). Routledge.

Hidayatullah, A., Csíkos, C., & Syarifuddin, S. (2023). Beliefs in mathematics learning and utility value as predictors of mathematics engagement among primary education students: The mediating role of self-efficacy. *Education, 3–13*, 1–14. https://doi.org/10.1080/03004279.2023.2294141

Hill, H. C., Ball, D. L., & Schilling, S. G. (2008). Unpacking pedagogical content knowledge: Conceptualizing and measuring teachers' topic-specific knowledge of students. *Journal for Research in Mathematics Education, 39*, 372–400.

Hutchins, P. (1968). *Rosie's walk*. Simon & Schuster.

Kozlowski, J., & Chamberlin, A. (2021). *Creativity-based mathematical instruction: Fostering a holistically creative mathematical environment*. Routledge.

Liu, N., & Twomey Fosnot, C. (2008). *The double decker bus*. Heinemann Educational Books.

Mackenzie, T. (2018). *The power of the provocation*. https://www.trevormackenzie.com/posts/2018/2/13/the-power-of-the-provocation

Marr, B. (2018). *How much data do we create every day? The mind-blowing stats everyone should read*. https://www.forbes.com/sites/bernardmarr/2018/05/21/how-much-data-do-we-create-every-day-the-mind-blowing-stats-everyone-should-read/?sh=4ede47dc60ba

Mehta, R., Mishra, P., & Henriksen, D. (2016). Creativity in mathematics and beyond – Learning from Fields medal winners. *Tech Trends, 60*(1), 14–18. https://doi.org/10.1007/s11528-015-0011-6

Melfi, T. (Director). (2016). *Hidden figures* [Film]. Twentieth Century Fox.

Mindset Works. (2017). *The impact of a growth mindset: Why do mindsets matter?* https://www.mindsetworks.com/science/Impact

Murtagh, E. M., Sawalma, J., & Martin, R. (2022). Playful maths! The influence of play-based learning on academic performance of Palestinian primary school children. *Educational Research for Policy and Practice, 21*(3), 407–26. https://doi.org/10.1007/s10671-022-09312-5

Neuschwander, C. (2009). *Sir Cumference and all the king's tens: A math adventure* (Illustrated by W. Hunsberger). Charlesbridge.

Newman, C., & Langbroek, J-P. (2014). *Great results guaranteed for students in 2015.* https://statements.qld.gov.au/statements/75312

Novita, R., & Putra, M. (2016). Using task like PISA's problem to support student's creativity in mathematics. *Journal on Mathematics Education, 7*(1), 31–42.

NRICH, University of Cambridge. (2019). *Low threshold high ceiling – an introduction.* https://nrich.maths.org/10345

OECD. (2023a). *PISA 2022 results (Volume I): The state of learning and equity in education.* OECD Publishing. https://doi.org/10.1787/53f23881-en

OECD. (2023b). *PISA 2022 results (Volume I and II) – Country notes: Australia.* https://www.oecd.org/en/publications/pisa-2022-results-volume-i-and-ii-country-notes_ed6fbcc5-en/australia_e9346d47-en.html

OECD. (2024). *PISA Results 2022 (Volume III): Factsheets: Australia.* OECD Publishing. https://www.oecd.org/en/publications/pisa-results-2022-volume-iii-factsheets_041a90f1-en/australia_167b4caf-en.html

Paatsch, L., Casey, S., Green, A., & Stagnitti, K. (2024). *Learning through play in the primary school: The why and the how for teachers and school leaders.* Routledge. https://doi.org/10.4324/9781003296782

Parlakian, R. (2022). Nurturing early math play and discovery. *Young Children, 77*(3), 88–92.

Piggot, J. (2017). *Bloom's taxonomy.* https://nrich.maths.org/5826

Ramani, G. B., & Eason, S. H. (2015). It all adds up: Learning early math through play and games. *Phi Delta Kappan, 96*(8), 27–32. https://doi.org/10.1177/0031721715583959

Ribeiro, A. R., Pereira, A. I., Pedro, M., & Roberto, M. S. (2023). Predictors of child student engagement in elementary school: A mixed-methods study exploring the role of externalising problems. *Infant and Child Development, 32*(5), e2449. https://doi.org/10.1002/icd.2449

Ritchie, S. J., & Bates, T. C. (2013). Enduring links from childhood mathematics and reading achievement to adult socioeconomic status. *Psychological Science, 24*(7), 1301–8. https://doi.org/10.1177/0956797612466268

Shulman, L. S. (1986). Those who understand: Knowledge growth in teaching. *American Educational Research Journal, 15*(2), 6–10.

Sousa, D. (2015). *How the brain learns mathematics* (2nd ed.). Corwin.

The Education Hub. (2021). *Piaget's theory of education.* https://theeducationhub.org.nz/piagets-theory-of-education/

The Sydney Children's Hospital Network. (2025). *Children (5–12 years): Creating a supportive environment for children's healthy growth and development.* https://www.schn.health.nsw.gov.au/kids-health-hub/child-development/child-development-children-5-12-years

Victorian State Government Department of Education. (2024). *Numeracy for all learners.* https://www.education.vic.gov.au/school/teachers/teachingresources/discipline/maths/Pages/numeracy-for-all-learners.aspx

World Economic Forum. (2025). *The future of jobs report 2025.* https://www.weforum.org/reports/the-future-of-jobs-report-2025/

Yeager, D., & Dweck, C. (2020) What can be learned from growth mindset controversies? *The American Psychologist, 75*(9), 1269–84. https://doi.org/10.1037/amp0000794

www.ingramcontent.com/pod-product-compliance
Lightning Source LLC
Chambersburg PA
CBHW052210090526
44584CB00016BA/1999